COMBAT ENGINEER BRIDGE SPECIALIST

By

Edward Leo Semler Jr.

Copyright © 2023 by Edward Leo Semler Jr.

All rights reserved by the author.

First Edition: 2023

Library of Congress Control Number: 2023923354

ISBN: 978-1-737-6472-4-9

Printed in the United States of America

City of Publication: Schulenburg, Texas

Cover layout by Edward Leo Semler Jr. & picture from "Snap"

"To all my Army buddies"

TABLE OF CONTENTS

Introduction	1
Getting Ready To Join The Military	3
Boot Camp - Fort Leonard Wood Missouri	13
2nd Infantry Division Korea	47
5th Infantry Division Fort Polk Louisiana	139
After The Army	185
About the Author	189

INTRODUCTION

If you read my book "Around the Word," which is my memoirs, this book may seem very familiar. Because this book covers the same period of time, at the beginning of that book, while I was serving in the United States Army. But I felt that my time in the Army was not being represented that well in a book that seemed to portray the memoirs of a Coast Guardsman, and not someone that had spent time in the Army. I think readers were surprised to open the book and find that there were 113 pages about the Army before they even reached the Coast Guard.

This book tries to remedy that. It's a bit more detailed than what you'll read in "Around the World," has additional stories, and expands on several others. Since it was written 10 years after "Around the World," it also has updated information. Plus, I'm able to insert more pictures to help the reader visualize the stories I am telling.

In writing "Around the World" I didn't want to mention or expand on every story I had to tell, just the major milestones. This was due to the long amount of time I had spent in military service, over 25 years, and I had a lot of stories and time to cover. I didn't want to slow down the pace of that book with every detail in my career. Because who really wants to read a 600-page book!

And I'm really glad to have my Army time stand alone in its own book. In "Around the World" my time in the Army takes up a third of the book, which is a huge portion for just three years out of a 25-plus year career. But in that short three-year enlistment, I did a lot!

Anyone who has served in the Army, or any branch of service for that matter, will be able to relate to the stories I am about to tell. Because life in military service seems to be basically the same in all branches. And I can attest to this after serving in the Army and Coast Guard. The uniform and job description may change, but interacting with your fellow service members and overall mission to defend your country doesn't change.

GETTING READY TO JOIN THE MILITARY IN 1981

Joining the military started out at a young age and with a very simply desire to follow in my father's footsteps and join the United States Army. He had served in a peace-time Army from 1959 to 1962, mostly in France. He had joined as a cryptographer, a specialty dealing with codes, such as morse code. It was a very specialized job which required a lot of training. After the Army my dad got out and was offered a very nice job with the United States Government. It seemed so easy. And that was exactly what I wanted to do. Join the military, get out and be offered a great job, and raise a family.

My dad was a great storyteller, and he always knew when to embellish his stories to bring them to life, especially the stories about his time in the Army. He made his time as a cryptographer seem so much more exciting than I imagine it actually was. They intrigued me and I dreamed of growing up and experiencing exciting adventures like his.

As I grew older, I learned that there was more to military service then just the Army. There was the Air Force, Navy, Marines Corps, and the Coast Guard.

Once I had been flipping through a National Geographic magazine and read a very interesting article about the Coast

Guard. I asked my dad what he thought about joining them. He said that you needed a college degree, so that was quickly dashed from possible consideration.

There wasn't anyone in my immediate family who was currently in military service, so I didn't have a whole lot to go on. The Vietnam War and the many years of drafting men into service was still fresh in everyone's minds, so there wasn't a rush to enlist.

My extended family had a long history of service, with several uncles serving in WWII, my great-great grandfather and his two brothers serving with the Union Army during the Civil War, and my six-time great grandfather serving in the Revolutionary War. But they were all silent about their service or had passed away.

Because my dad worked for the government, we lived in various countries and on U.S. military bases. Here I loosely associated with military members and their families, mostly Air Force. The bases and support structure for military members seemed very appealing, with nice base housing and Exchanges to shop in. They always seemed well taken care of and had the financial means to drive nice cars and play endless rounds of golf.

As I grew into my teens, I played on softball teams and golfed with military members, and it seemed like such an inviting lifestyle. Granted I was only seeing the relaxing off-duty side of being in the military.

My parents had moved to Australia in 1979 where my father's employer, the United States Central Intelligence Agency, transferred him from Fairfax, Virginia to Alice Springs, Australia.

While there as a 16- & 17-year-old I engaged with members of the Air Forces Detachment 421. They ran a small facility there and handled our mail and Exchange orders. But the Air Force didn't appeal to me. They didn't seem military enough. And I didn't see my dad's adventurous stories playing out in their daily routine.

I attended high school in Australia until 1981 and then left after completing the 11th grade. My parents thought it would be in my best interest to send me back to the states to finish out my 12th grade year, instead of completing it in Australia. It wasn't that the Australian school system was inferior; it was that my parents thought an American High School diploma would translate better in the American workforce.

It was arranged that I would head back to the United States and live with my Grandma Semler and Uncle Mike, one of my dad's brothers, on the Semler family farm. Primarily because they had a spare room. The Semler Farm was situated in rural Gibsonia, Pennsylvania.

My actual legal guardians while living in the states were my Uncle Joe, my dad's brother, and his wife, my Aunt Gloria. They lived about 400 yards next door to the farm. At the time they had Aunt Gloria's parents living with them and they were tight on room. My grandma had a spare room at the farm, so it just seemed easier to stay with her and Uncle Mike.

When I went to Deer Lakes High School to register for my 12th grade year I immediately ran into problems with my Australian schooling. Deer Lakes said my high school credits from Australia would not give me enough credits to graduate after completing the 12th grade in the United States and I would have to complete the 11th grade again in the American school system. That was definitely not appealing to me. I was a mediocre student at best. And doing another year was unthinkable.

Luckily my Cousin Sharon, Uncle Joe and Aunt Gloria's daughter was a teacher in the school district. She was somehow able to demonstrate that I met the requirements for credits to complete the 12th grade and graduate with a diploma. What a relief!

I had immediate family all over the area and spent a lot of time in Gibsonia, West Deer, New Kensington, and greater Pittsburgh. In February of 1982, while in the middle of my senior year, I drove down to New Kensington to visit my Uncle Harry's family. While I was there, I decided to go see the Army recruiter in downtown New Kensington. As I walked into the recruiting building, I never gave a second thought to the offices of the Navy, Air Force or Marine Corps as I made my way straight to the Army Recruiting office.

When I arrived, there was an "out to lunch" sign on the Army Recruiters door. So, I patiently waited for their return. The next office down was the Marine Corps recruiter. As I waited, I started to pace up and down the hallway, and drew their attention. Out came a Marine looking to meet his recruiting goal. He started to chat me up and I told him I was waiting for the Army Recruiter. That comment just brushed right by him and he asked me "how many pull-ups can you do?" I said I don't know." Conveniently placed in the entrance frame to their door was a pull-up bar. The Marine said, "hop up there and knock me out at least 5." Which, at the time was a requirement to get into the Marines. I leaped up and grabbed the bar and hung there for a second. Then I strained to pull my body up to the bar and got one pull-up. That was it. I was done with just one-half ass pull-up. I wasn't impressed and neither was the Marine. I guess my smoking habit was catching up to me. The Marine then grudgingly left me to wait for the Army Recruiter.

When the Army Recruiters came back from lunch, they were happy to see me waiting and ushered me right in. I was assigned to a young sergeant (SGT) by the name of Ernest Roland, who like any recruiter was very eager to get me signed up and meet his hiring quota. And he didn't even have to go out and drag me off the street; I was a willing applicant that walked right up to his desk! He was a thinly built black man not to much older than me and was probably still on his first 3 or 4 year hitch.

After we started to fill out the required paperwork, SGT Roland seemed concerned. When he asked where I had lived and worked the past few years, and I said Australia, it seemed to throw a wrench in the Army's finely tuned recruiting machine. But after a lot of additional paperwork, because of my unique circumstance, I was signed up on the delayed enlistment program. If I passed all my pre-entry requirements I would ship out to basic training in July after I graduated from high school.

In preparation for my deployment in July I was routinely contacted by SGT Roland, who would drive out to the farm and check in on me. On one occasion he came out in his government vehicle, a tiny little Dodge Horizon. The lane up to the farm was about 200 yards long and the worst stretch of driveway in the township. When that lane was wet it was like driving through mini lake-sized potholes all the way up to the house. You actually had to steer clear of them for fear of getting stuck in one. Well, SGT Roland came out after a good rain storm and got that little Dodge Horizon stuck in the lane. Luckily, we had a few old farming tractors there on the farm and I started one up and pulled him out.

I had to make several trips into the Pittsburgh Military Entrance Processing Station (MEPS) located at the Federal Building in downtown Pittsburgh. It was about an hour's drive and pretty much a straight shot down the main drag of Route 8. The trips consisted of the usual pre-entry physicals and an entrance test known as the Armed Services Vocational Aptitude Battery (ASVAB). The ASVAB was an eclectic mixture of subjects to try and find out what career path was best suited for you in the military. Each service has a different minimum score requirement to gain entry into their branch of the service. I passed the Army's required score and was one step closer to shipping out.

When at MEPS I would usually look up my Aunt Betty who worked for the Internal Revenue Service in the same building. I would either visit her office or meet her for lunch down in the cafeteria. She had such an upbeat personality that always put a positive spin on the visits.

On one of these visits to MEPS I had to decide what my job field would be in the Army. The female Army counselor who was assigned to me sat me down and asked me what I wanted to do. My reply was, "I don't know. What does the Army need?" Now I was putting a lot of trust in her hands! She said "How about being a combat bridge crewman, your ASVAB general technical (GT) score of 100 shows you're pretty strong in this job field." And she followed that up with, "I can even get you a guaranteed assignment to Korea." Now I thought this all sounded just great. This had adventure written all over it. Little did I know that a combat engineer was just one step up from an infantryman and no one wanted to go to Korea!

On another trip to MEPS I had to take my physical. This is an all-day affair going from one room to another and waiting for your turn to be seen to give blood, have your eyes checked, provide a urine sample, and so on. At one point in the physical they piled the 40 or so of us military candidates into a room and told us to line up in rows and get undressed down to our underwear. After complying with the command, a doctor with his assistant walked up and down the rows. As the doctor made comments, his assistant took notes. The doctor had walked past the front of me without much fuss and continued down the row. I heard him walking up the row behind me, heard him talking, and then felt his cold hands on my back. He told me I had a curvature of the spine and would need to step out of the formation and wait for him in the next room. After the doctor had finished his group exam he came into the room and looked me over some more. He said he was concerned about my curved spine and the pressures that would be put on it from carrying a back pack - ruck sack in Army lingo. He wanted to have me re-evaluated by another doctor the following week.

At this point I thought my military career had come to an abrupt end. I went home feeling rejected. After getting back home I decided to go visit my Grandpap and Grandma Churilla who lived about 10 minutes from the Semler Farm in neighboring Hampton. I had visited them on a regular basis while attending school and thought they may be able to give me some advice. My Grandma said that she had a chiropractor that she saw on a regular basis and recommended that I see him and get a back adjustment. I had always thought of chiropractors as quacks and didn't know what this guy could offer me. My Grandma reassured me and sweetened the deal by offering to pay for the visit. Being a cash-strapped young teenager, that was an offer I could not refuse.

So, we drove over to the chiropractor and he assured me he could get my back into alignment. He said the adjustment should last about 30 days or so and I should see him for regular adjustment to keep it straight. To accomplish the adjustment, he got me on his table and did the usual cracking, pulling, stretching, and low and behold I was as good as new and felt great. I felt very optimistic about my upcoming meeting with the MEPS doctor. As scheduled, I went to MEPS the following week, checked in with Aunt Betty, and was examined by the second MEPS doctor. He said he couldn't find anything the matter with my back and marked me fit for duty in the Army!

On my last night before shipping out to basic training my Uncle Mike and Cousin Dave threw me a party at the farm. We had a mini keg of Iron City Light beer and some pot, so we got drunk and stoned. Beer and pot were our two major food groups and we totally over-consumed that night. I remember Dave leaving the barn we were partying in and heading back to the house. A few hours later Mike and I decided to turn in and we found Dave passed out in the driveway, halfway between the barn and the house. It was a warm July evening, so we just left him there. I'm sure we were too messed up to be carrying him anyway.

Needless to say, the next morning was rough. I was hung over and don't think I had but 2 hours of sleep when SGT Roland arrived to take me to MEPS.

On the way down the lane, I was still pretty drunk from the night before. SGT Roland seemed a little nervous and he began telling me that he knew he had said a lot of great things about the Army, but boot camp was going to be tough, tougher than he may have let on. I looked at him with an unconcerned expression and told him I kind of knew that and not to worry, I would be fine. My only thought at the moment was that I just wanted my head to stop pounding and to take a nap.

At MEPS everything seemed to sort of move in a fog. We finished up paperwork, swore our allegiance to the United States of America and then headed to the airport for our flight to St. Louis, Missouri. It was the 21st of July 1982, and I was headed to Fort Leonard Wood, Missouri for basic training, also known as "boot camp."

UNITED STATES ARMY
FORT LEONARD WOOD, MISSOURI
2nd PLATOON, B COMPANY, 1ST BATTALION,
2nd TRAINING BRIGADE
21 July 1982 – 3 November 1982

It was late Wednesday, the 21st of July, and the plane touched down in St. Louis without me noticing. I was awakened by a stewardess nudging me who nicely said it was time for me to get off the plane. As I lifted my head, I could feel the bad crick in my neck and noticed that I had drool trickling from my mouth onto my lap. I felt like crap. I remembered Dave passed out in the driveway and believe it or not wished I was him right about then. At least he wasn't heading off to the Army. I got off the plane and headed to the mustering point to catch the bus to Fort Leonard Wood. I could tell I was getting closer by the number of people I saw carrying large manila envelopes. We were given these envelopes to carry our records to boot camp in, a dead giveaway you were a recruit.

After the bus filled up it headed into the middle of Missouri, about two and half hours from St. Louis. Fort Leonard Wood was out by itself with nothing much around it. I think the military likes it that way so if you decide to run away you really don't have any place close to run to.

Entering Fort Leonard Wood

The fort was named in honor of Major General Leonard Wood, who was awarded the United States Army Medal of Honor for his actions fighting against Geronimo – a native American Indian. He also saw action in the Spanish American War, Philippine American War, and World War I. He was a physician by training and as a neat historical fact he is buried in Arlington National Cemetery, but his brain is held at Yale

University. Of course, I knew none of that at the time and it probably would not have made much of an impression on my 18-year-old mind in any case.

The bus ride was uneventful, and I dozed on and off until we rolled into the fort late in the evening well after dark. When the bus came to a stop some average looking Army folks mustered us all off the bus and into an old wooden building for a quick briefing, basically stating that we were in fact at Fort Leonard Wood. In the building, which was one large room, there was a wooden box with a small opening on the top. This was near the door, and we were told that if we had anything we shouldn't have on us, that this was our chance for amnesty and, to put it in the box as we left. I had nothing to put in the box and moved on with the rest of the group to the next old WWII style two story building where we were provided a bed and were told to turn in. Which at this point I was eager to do.

Initially, I was surprised at how relaxed things were here. I had heard that basic training was tough, but that didn't seem to be the case. No one had told me all the fine details about boot camp. My dad just said, "It was no picnic" and SGT Rowland said, "It may be tougher than I have let on." The only traumatic event to happen the first week was that first night.

The sergeant that had tucked us all in for the night had assigned several of us to stand what he called a "fire watch." He assigned different hours of watch throughout the night to different people and departed with one final order, "Do not under any circumstances let anyone in until I get here in the morning." I was assigned to a bunk bed on the second floor. Finally sober from the night before I fell fast asleep.

At some point in the night, I was awakened by the sound of heavy pounding on the wooden barracks door and the commotion of people arguing. As I sprang out of my bunk and came down the stairs, I encountered about 10 guys huddled by the front wooden door. They looked scared out of their wits and were arguing the point of letting this guy pounding on the door in. The guy at the door was pounding and yelling to let him in or he would throw all of us out of the Army, or worse! Meanwhile, the group of now about 20 of us were huddled inside and having a delirious debate on whether or not to let this guy in. There was screaming and crying and all sorts of carrying on and eventually the unknown person at the door was let in.

Needless to say, it was a no-win situation. The unknown guy came barging in with a female on his side and a couple of male sidekicks in tow, all of whom were in plain green uniform. They seemed to be about our young age of 18, and more than likely just got here a few weeks before us! The female seemed to be the ring leader's date and she had hickeys all around her neck. The ringleader gave us an ass chewing for letting him in and triumphantly strolled out the door with his girl and sidekicks in tow, leaving us all petrified.

All the commotion had everyone worked up and we all thought we were in big trouble and were going to be sent back home. But the next day came and went without a mention of the evening event. Nothing was ever said about it. I think someone just got their jollies by whipping the new recruits into a hysterical frenzy.

I later found out that we were just at a holding and processing center, and not boot camp. Although there were female recruits and cadre at Fort Leonard Wood, I was in an all-male forming company. Our days were filled with getting issued uniforms, gear, dog tags, pictures, haircuts and equipment. This always entailed about 30-40 of us sort of marching to a different WWII era building and spending hours awaiting important items such as underwear, which was the only thing I had a choice in; boxers or briefs.

It was a sort of casual routine. We got up when revile was announced and made our way to the mess hall, which was a huge tent structure that seemed sort of temporary. It had a wooden floor with wooden walls that went up about three feet and stopped. The rest was canvas tenting material. The side walls rolled up and you had a nice view of the processing area while you ate. After breakfast we would form up and go through the day of in-processing.

Getting our patriotic picture taken in uniform was another interesting event in this finely tuned process. At this point, we were all still wearing the civilian clothes we had arrived in. You entered the building and there was a rack of Army dress shirts and a pile of dress hats on a table.

You picked an Army shirt and hat that fit you and put it on. You sat very officially in front of the camera and got your picture taken. Upon completion of the picture, you removed your Army shirt and hat, placed it on the rack and table where you found it, and exited to wait for everyone else to finish.

My basic training picture

A few days later we were issued our real uniforms, and our civilian clothes were boxed up and shipped back home. There would be no need for them in the next few months. My group was one of the first to be issued the Army's new battle dress uniform (BDU) which was a camouflage pattern. The older uniform was just plain green. I was really happy with the new BDUs because in my opinion they made you look more like a soldier.

After spending five leisurely days at the in-processing center we started packing up our gear for the inevitable. And on Monday the 26th of July we were thrown to the wolves!

Transfer day started out early in the morning with us loading everything we had into our rucksack and duffle bag. We had been told that we should only have what was issued to us at the processing center, nothing else. I don't think a lot of guys let that sink in, because some guys had a hard time getting everything to fit.

We were mustered in front of our old wooden barracks with all of our gear. We looked like we were shipping out to the front lines, wearing our BDUs with metal helmets. We were told that we would be picked up by cattle cars that would transit us to our new living quarters. Now these cattle cars were just that, real looking cattle cars towed by a semi-looking truck. There were no seats, just handrails to grab on to. No windows, just slits in the metal walls with bars on them. The only way in or out was through a set of double school bus style doors in the middle of the cattle car on one side. We all loaded up as tight as sardines into the arriving cattle cars. As I got on, I noticed there were several guys wearing the older plain green uniforms and wide brimmed, brown, "Smokey the Bear" type hats. These guys didn't say a word and they didn't look happy. I could sense that bad things were about to happen.

As we started to pull away a small amount of chatter started up. As soon as the chatter started the guys in the Smokey Bear hats yelled "Shut your fucking mouths!" Well, there was a deathly silence as we drove around for what seemed like an hour. I guess they didn't want us to know where we were going or how to get back to where we had come from. Finally, our cattle car pulled up to this long and wide concrete driveway leading up to sets of newer style brick buildings, each being three stories high. As soon as the cattle car stopped, the school bus doors sprang open, and those guys with the Smokey Bear hats just started yelling and screaming and pushing guys off the cattle car. I mean they were just grabbing whoever was close and throwing them right out the door to even more guys wearing those Smokey Bear hats.

I was toward the back of the cattle car, which seemed to be a good place for the moment, as the Smokey Bears were busy with everyone at the front. As the car emptied and I made my way to the doors I hit the ground a-running and headed up the long grade of the concrete path. I had my rucksack on my back and my duffle bag in front of me in a bear hug. It was like running a gauntlet. There were guys on the ground crying and screaming, their gear strewn about creating obstacles in my path. It was like what you see in the war movies, when the landing craft reaches the beach and the front ramp drops down, and there is a full-blown war going on right there in front of you. And the Smokey Bears yelling and screaming sounded like incoming artillery. I didn't really know what they wanted me to do, but the trail of destruction seemed to be leading up the hill, about 100 yards and in front of one of the three story buildings. I made it to the front of the building and fell into formation with those who had survived the gauntlet ahead of me.

We were standing at attention in two rows facing each other about 10 feet apart. There was about five feet in between each man and in the middle of the two rows of men there were trash cans. The Smokey Bears informed us that they were to be addressed as "drill sergeant" and we were "maggots."

They then began to dump out everything we had packed in our duffle bag and rucksack onto the hot July concrete. As they dumped my gear and clothes on the concrete, the drill sergeant started to kick it all around looking for anything he didn't think I should have. He used his feet to sift through my gear giving me the impression he was above actually touching it with his hands. Once he was satisfied that I didn't have anything interesting, I was told to pick up my shit. As I was scrambling to get all my belongings back into some sort of pile, I was thankful I wasn't like the guy next to me who had magazines and candy, an obvious no- no! The drill sergeant was throwing this guy's stuff all over the place and making him do push-ups for all his contraband. Another guy had pantyhose. I thought, what in the heck was this guy thinking! His rant to the screaming drill sergeant was that they are the best way to put a shine on his boots. The drill sergeant was not impressed, threw the panty hose in the garbage can, and made the guy feel miserable for a while.

There were also these two Hispanic kids that looked like "deer caught in headlights." They drew the attention of the head drill instructor who appeared Hispanic himself. He was yelling at them, asking them if they understood English. They really didn't seem to understand, and like the rest of us seemed too afraid to talk. They were pulled aside, and I asked myself, how the hell did these guys get this far in this process not knowing how to speak English?

The drill sergeants, now content with our total humiliation, had us run up to the 3rd floor of the building, which would become our new home for the next three plus months.

I was in a room with about 20 other guys. It was an open room with metal lockers against the walls and two rows of metal bunk beds in the center. I was on a top bunk and this would be my home for the next 14 weeks, from the 26th of July to the 3rd of November.

The barracks

We must have been one of the first waves to arrive as our company formed up. Because over the next several days we watched as other cattle cars pulled up to the concrete welcoming mat just down the hill and the same scenario played itself out with newly arriving maggots.

I was on a working detail the next day and we were told to take a break out of the heat under a tree in front of the building. The heat waves radiating from the concrete pad in front of the barracks reminded me of being out in the Australian desert. After sitting there for a few minutes, a group of cattle cars pulled up below us and I had a front row seat for the main event. It played out just like our arrival the day before. But once the drill sergeants had all the maggots up on the main concrete pad, a few of them passed out. One of the drill sergeants yelled at us to "Get your butts over here right now!" So, we sprang up and ran over to him. Once there we were instructed to pour our canteens of water over the dazed maggots laying there on the concrete and drag them under the tree we had been sitting under. It seemed to revive them. Once back on their feet the process of dumping their gear out and so on continued.

The next few weeks were jam packed with events from 0330 until 2200. That's military time for 3:30am-10pm. Each day began with all of us springing out of our bunks and trying to cram into the bathroom to shit, shower, and shave. I had nothing to shave but they said I had too anyway. Also, our bunks had to be made and everything cleaned up within 30 minutes after waking up. Then it was down to our mustering spot in front of our barracks and morning physical training (PT) consisting of jumping jacks, push-ups, stretches, and a good mile or so run. Once back from PT we headed to the mess hall for breakfast and on with our day.

There were about 211 of us in Bravo Company when we started. I would have to say we lost about 25, who were discharged or reverted back to another company by the time we finished. I was in second platoon, which consisted of about 72 when we started. My friends were the guys in my immediate platoon who I spent every minute of every day with. Guys like Knuckles, Vaughan, Kyle, and Sann. Everyone was known and addressed by their last names, mimicking how the drill sergeants addressed us. You never really got personally close with anyone because you never knew how long they would be around, and you were concerned foremost with getting yourself through this mess. If you didn't get kicked out you would be reverted back to another company behind us, and your boot camp experience would drag on.

It seemed like every week we would lose one or two guys for one reason or another. A few that stand out were our recruit company commander and a fellow in my own platoon. We were into our second or third week when we lost our recruit company commander. He was an older guy that I heard had been in the Army before. The drill sergeant came out one day and told him to go pack his gear because he was leaving.

2nd Platoon, Bravo Company, 1st Battalion, just after we formed up. SFC Castaneda is holding the flag. I'm the 7th from the left on the top row.

Word was that he falsified something on his enlistment papers and they had finally caught it. This was before computers and everything was hand generated, so it took time to verify your paperwork. At about that point I was wishing they would come and give me my walking papers. I remember standing there in formation about midway through our seven week basic training and thinking, there is no way I will be able to maintain this pace for my three year enlistment!

The other guy I remember leaving was a fella by the name of Taylor. He was a nice, reserved guy who had hygiene problems. We were always getting in trouble and having to do push-ups or put in pain inducing positions because Taylor would not bathe, or his gear was dirty.

One of the drill sergeant's favorite pain inducing positions was for you to stand up against the wall, bend your knees, and hold your arms out. It didn't take long before you started to shake, and it was everything you could do to keep your arms up. If you couldn't hold it for long, the drill sergeant would come over and poke you or give you his boot. You didn't want the attention of the drill sergeant.

We tried all the usual things to get Taylor to clean up his act, such as hazing him and throwing him in the shower. None of it seemed to work and he really didn't seem to care. Well one day we were at the gun range shooting our M16 rifles. We were all sitting Indian style in this gravel parking lot waiting our turn to fire when we noticed that Taylor was shoving pebbles from the parking lot down the barrel of his rifle. One of the guys notified our drill sergeant and Taylor was whisked away, and we never saw Taylor again.

Besides a pulled tendon and a bad rash around my neck from my dog tags I made it through without a scratch. No one wanted to pull up lame for fear of being reverted back to another company or discharged. We ran and did all of our physical training in boots, BDU pants, and t-shirts. The boots were your typical black military boot that had no support, and were about 13 inches tall. After several weeks my Achilles tendon gave out on me. I tried to keep up the pace, but the drill sergeants picked me out limping along and sent me to the aid station to have it looked at. At the aid station they gave me some pads and put me on light activity for a few days. Of course that got the drill sergeants all over me, accusing me of being a slacker.

The guys that were sick, lame, or lazy, were like a leper colony. Whenever we formed up to march some place they always formed up behind the company and walked in a group trailing along behind, so they didn't infect the rest of the company. They were easy pickings for the drill sergeants and usually didn't last long before being reverted back to another company or discharged. I spent a day or two back with them with my tendon issue and was very happy to get out of their ranks and back up with the company.

In letters home I was proud to report that I had quit smoking, was up to 44 push-ups, 49 sit-ups, and that I could run the mile in 7.48 minutes. The requirements were 40 push-ups and sit–ups within two minutes, and to make the run in less than 8 minutes, so I was passing. I earned sharpshooter with the M16 rifle by hitting 34 out of 40 and marksman with the grenade. I'm not sure why they have an award level for grenade. You would think it would be simple, like pass or fail. Either you got the grenade close enough to inflict damage or not. No matter what the scoring was based on, the grenade course and fire and maneuver ranges were cool, and I felt like I was learning something important.

Before completing the grenade course we had to complete the grenade range. This involved throwing a bunch of training grenades to get the hang of having the explosive in your hand. The training grenade was a blue colored grenade with a blasting cap in it to make the sound of a real grenade. It really gave you the feel for the weight of it and the time delay before the explosive went off.

At the practice grenade range

Next it was on to the actual grenade course. This was about the length of two football fields and had a series of bunkers, foxholes, and vehicles that you had to sneak up on and then throw your training grenade at. It seems easy until you crawl about as close as you can get to the obstacle wearing all your combat gear and then try and lob a grenade into the small opening of your target. On some of the obstacles you had to make sure the grenade exploded just after landing. This was to make sure the enemy didn't have time to pick it up and throw it back at you. You did this by popping the pin on the grenade, waiting a few seconds and then throwing it. Once you popped the pin, the grenade was supposed to explode in eight seconds, so you had to calculate the wait time according to the distance to your target. After about halfway through the course I was beat! My adrenalin was pumping and it felt like a real accomplishment to finish. I was scored on how many times I got my grenade on or in the target. I thought I done pretty well, but ended up earning a marksman medal, which was the minimum standard for qualifying.

Finally, it was on to the last part of the grenade training which was to throw a real grenade! We were mustered up in the usual parking lot sized holding area and called up to the grenade range bunker in groups of five.

When I walked into the bunker, I was given a flak jacket to wear along with instructions on how I would handle and throw the grenade. When it was my group's turn, we were each handed a grenade and proceeded to five small individual bunkers. I was in the third bunker. The back side of the bunker was open to provide entry and the front and sides were piled five feet high with sandbags and wooden beams. There was a drill instructor there and he instructed me to crouch down, and he once again went over the instruction for throwing the grenade.

As I heard the first position throw his grenade, I started to get nervous. When the second position threw his I was even more on edge. It was my turn now.

The drill instructor gave me the order to stand, pull the pin on the grenade, and hand the pin to him, which I did. I now had a live grenade in my hand. I'm not sure who was more concerned, me or the drill instructor.

Following his orders I yelled, "Fire in the hole!" and threw the grenade down range and dropped to my knees. A few seconds later there was a loud explosion. What a relief!

It was the fourth position's turn and after waiting several minutes for an explosion there was nothing but silence. It was a dud. I was told to stay put and I waited there in the bunker with the drill instructor crouched down against the sandbags. After about 45 minutes an explosive ordinance team was called in to dispose of the dud grenade. They calmly walked down range and looked around for the dud. Once they found it, they wired it with explosives. Then they walked back to the bunker area and detonated it, exploding the dud.

The fifth position threw his grenade without a hitch, and I was glad to get out of that bunker and back out to the safe holding area!

The fire and maneuver course was pretty much the same as the grenade course, only I used my M16 rifle instead of the training grenades. I had blank rounds in my weapon along with a partner who went through the course with me. Our objective was to advance on a target while giving each other covering fire. You would lose points if you moved before you had covering fire or did not provide covering fire when your partner moved. No medal for this, just a pass or fail.

There were several other ranges such as the M203 grenade launcher and the M72 Light Anti-Tank Weapon (LAW) rocket launcher. These ranges were fun, but we never fired the real thing. At the M203 grenade launcher range we fired these dummy rounds that were filled with paint. The M203 attaches to the underside of the M16 rifle, and we would line up and get one shot at firing the thing down range at a window sized opening in a wooden panel.

At the M72 LAW range we fired a dummy round that was similar to a bottle rocket. I'm sure it was to save money and I know that if I was a drill sergeant I would sure prefer teaching us on weapons that would cause little damage if we made a mistake.

The company drill sergeants and drill assistants were an eclectic lot of Hispanics, African Americans, and Caucasians. Our platoon's lead drill sergeant was Sergeant First Class (SFC) Juan E. Castaneda. He was our primary drill sergeant, but any one of the company drill sergeants or assistants could be in charge of us at any given time.

SFC Castaneda was serious and stern. I remember always having a feeling of trust with him, knowing that he would not let anything bad actually happen to us. I saw his compassionate side that first day with the Hispanic guys that could barely speak English. Although he didn't let them off the hook with going through the first day welcoming party, you could tell that

SFC Juan E. Castaneda

he was not giving them the full treatment once he ascertained that they didn't get it.

Speaking of those Hispanic recruits, the last time I saw them was at the gas chamber. This was to simulate a chemical attack or exposure. As we were being prepared to enter the gas chamber, these guys didn't comprehend what was going to happen. The gas chamber was a sealed room filled with some sort of tear gas. We were told that we would enter the gas chamber in small groups of 10 with our gas masks and chemical suits on. When we were all inside we would be told to remove our masks. Once we yelled out our full name and social security number we would be allowed to leave.

When you removed your mask, it didn't take but a split second to be overcome with tear gas, your eyes burning and gushing with water. The instructions must have been lost in translation with these two Hispanic fellas. They came out of that gas chamber running and screaming as if someone were truly trying to kill them. They had no idea they were going to get gassed. By the time the drill sergeants got them under control, by tackling them, they were still freaked out. They were in such bad shape an ambulance was called to take them away. I think after that they were reverted back to another company a few weeks behind us to be retrained, and most likely enrolled in English classes.

The other drill sergeants in the company that assisted SFC Castaneda interacted with us daily. Very seldom did we do things as just a platoon. Just about every day we functioned as a company and any one of these drill sergeant could be in charge of us. The drill sergeants were SGT Abraham, Staff Sergeant (SSG) Miller, SSG Strickland, SFC Chatter, SFC Wiley, and SGT Burrell. There were usually two or three of them around during the day, but it was very unusual to have them all together on one given day, but make no mistake, one was always around. Most of the daily barrack's routine such as cleanups, watch, and getting formed up was left up to our recruit leaders. The drill sergeants and their assistants would patrol through at various times of the day and night and crank us up, just to keep our heads on a swivel.

Drill Instructors

Once we were formed outside, we fell into the hands of the drill sergeants and their assistants. The drill sergeant assistants were a sort of rotating lot of drill sergeants in training. The daily barometer of how hard the day would be seemed to be measured by how many drill sergeants walked through the door in the morning to get us cranked up and on with our daily routine. They all had us under total control and could break us at any given moment. Some just rode us hard for a while and backed off once we got going, while others gave it to us nonstop the whole day.

After a few weeks we were marched down to the mini–Post Exchange (PX) that was in our training area. The large main PX is a store similar to a department store. The mini PX was like a convenience store. We had been issued an advance of $90.00 when we began training and told to budget this out for haircuts and toiletries. I was thinking what the heck else are we going to spend it on?

When we arrived at the mini PX the drill sergeant told us "You have 30 minutes to get in there and get a haircut, toiletries, and get back out here in formation!" Most of us double timed it in there, got our head shaved, got our toiletries, and got back out. Of course, there were a couple of guys that came out with a standard haircut, which did not impress the drill sergeant who promptly called them pretty boys and sent them back in to get it shaved off!

Some other memorable moments were the first time we received a pass, the "unlocked locker" fears, the mess hall routine and the "sticky bun" matter, mail call, and my mattress turn in.

We received our first pass after the first few weeks of basic training known as "total control." It's called total control because the drill sergeants demonstrate they have total control over you by breaking you down and maintaining total control of your every movement. After they were satisfied that they had us scared to breathe, they let us go on pass for a few hours. We were only allowed to go a couple of blocks away from our barracks but it was freedom!

A bunch of us went over to the Davis Club, which was a huge building especially made to accommodate a large number of recruits at one time for drinking and socializing while on pass. I had two beers. I wanted more but knew better. The two hit me pretty hard as it was. Watching my peers it's amazing how many beers a guy can put away in a few hours on pass. Overdoing it on pass was obviously expected by our drill sergeants and at the end of our pass they mustered us all out in front of the barracks and walked the ranks looking for anyone intoxicated. When they found one, they had them get out on the grass and had them doing push-ups, running in place, drop and rolling, and a wide variety of sobering exercises. They usually were not satisfied with your sobriety until you vomited all you had consumed out onto the grass. Lesson learned; don't drink too much while on pass!

The fear of leaving your locker unlocked was totally ingrained within a day or two of arriving at our new third story home. We were instructed to never, ever leave our lockers unlocked. After the second day we came marching down the maze of road sized concrete paths through endless barracks right up to the concrete pad in front of our barracks. As we did so there was a pile of someone's belongings; clothes, gear, and personal items strewn all over the concrete pad. We marched right through it as if it was not even there. Some poor guy's stuff had been thrown out the third story barracks window onto the ground below.

The drill sergeant stated that the owner of this gear had better get it picked up and fast! So, we broke formation and ran upstairs fearing for the worst. Luckily it was not my gear. However, this incident initiated my bad habit of checking and rechecking my locker about 20 times before I could turn my back and leave it! I unfortunately still have that habit to this day. It worked with the other guys as well because it was very rare that anyone's stuff was lying out on the concrete when we returned.

Eating meals at the mess hall was the highlight of the day because we basically got a chance to eat. I mean you got to eat at breakfast, lunch, and dinner and that was it. There is no snacking, and this was the only time you came into contact with food. The routine was simple and as usual with the Army there were no wasted minutes or steps.

The process started with us marching down to the mess hall singing our cadence. SGT Abraham was an excellent cadence caller and my favorite. A good cadence caller picked just the right peppy tune to keep us in step and lift our spirits. It instilled a lot of pride in us when we had the tune down and managed to stay in step. Some of my favorites were, "nine to the front and six to the rear, that's the way we do it here," which told us how far our arms should move forward and backwards when marching. And "ain't no use in looking down, ain't no discharge on the ground," which told us to march with our heads up high.

Once we arrived at the mess hall, we would form two rows and have to go through a set of monkey bars about 20 feet long and 10 feet high. When you cleared the monkey bars you had to do 10 push-ups and 10 sit-ups followed by getting in line to enter the mess hall. If for some reason you fell from the monkey bars you got back in line to go through them again. Once in the mess hall your meal became a timed event. There was no choice in the matter, just take what was piled on your tray and get to a table as fast as you could and eat. Of course, it was milk, water, or juice for a beverage. That's not to say there were no soda machines. There were, but only for the drill sergeants. When the last man entered the mess hall the first man was to be heading out the door. The drill sergeants had a keen eye for how long you had been in there and you always had just enough time to get your food down before the cattle prod of the drill sergeants' eye had you up and moving.

On one occasion after leaving the mess hall we came marching up to the barracks and we did our usual facing movement towards the barracks. Our drill sergeant stated that he knew someone had taken a sticky bun from the mess hall and they were going to have an inspection to see who had it. All of a sudden, a sticky bun appeared from nowhere on the ground, in the middle of the formation.

Everyone wanted away from that sticky bun as if it were a live grenade! But of course, we were at the position of attention and could not move. The drill sergeants were not happy with this total disrespect for their authority. It was a long evening out there on the concrete pad getting drilled for someone taking that sticky bun. They wanted the sticky bun thief to come forward. But the thief never came forward that I can remember. And no one was going to tell on him. Because the infraction worse than getting caught with a sticky bun would have been snitching someone out for having the sticky bun!

Punishment always trickled down our recruit chain of command, from our company commander to the platoon leaders to the section leaders and so on. I was in no leadership position while in boot camp and maintained a very low profile in the back row of our formation. The higher you were in the recruit chain of command the harsher the punishment was from the drill sergeants. It was obvious to me after day one that I didn't want a position of authority or any type of visibility in boot camp. These drill sergeants had no problem putting their hands on you. They would grab you, throw you around, and call you every profanity they could think of. They did not want any misunderstanding about who was in charge.

Ranked right up there with chow was mail call. Nothing boosts a soldier's morale like letters from home. During mail call we would be in formation on our beloved concrete pad. If you had mail, the drill sergeant would call your name, you would have to break formation properly, run up, and get your letter. For me this was done by taking a step backwards, coming to the position of attention, pivoting to my right, marching to the end of the formation, pivoting left, sprinting to the drill sergeant at the head of the formation, stopping in front of him at the position of attention and stating, "Private Semler reporting for mail."

Nothing is free in boot camp and to receive your mail you had to pay for it with 10 push-ups. No need for the drill sergeant to wonder if you did all ten because you had to scream, "one drill sergeant, two drill sergeant," and so on at the top of your lungs while making your payment. And you didn't want to get a package! You had to open it up there in front of the drill sergeant so he could inspect it for contraband. And if you had contraband, you just lost it and gained push-ups.

Well, one day SSG Strickland was holding mail call and called "Semler." It was my first letter, and I came running up to get it. SSG Strickland stopped me before I dropped to pay for it in push-ups. With his smoke-glazed sun glasses that obscured his eyes he said, "Semler, I don't remember you, have you been hiding from me?" I said, "No drill sergeant, I have been here in the back row since the beginning." SSG Strickland said, "Semler you have been hiding from me, so give me 20." I snapped off with a loud, "Yes drill sergeant" and commenced my push-ups as my letter was tossed on the ground for me to collect when payment was made in full.

Two weeks before the end of basic training, over the Labor Day weekend of the 4th through the 6th of September, the drill sergeants solicited for 40 volunteers who still had money left over from our initial advance of $90.00 when we entered boot camp. I was a little hesitant to volunteer because I was warned by my dad to never volunteer for anything in the Army. Up to this point I had followed this rule, but for some reason decided to give in since I had plenty of my initial advancement of pay. Turns out it was a great move because they took us on a two-day pass to the Lake of the Ozarks state park. We loaded up on buses and made the hour drive to a beautiful retreat area that we had all to ourselves. Even though we were under the watchful eyes of our drill sergeant it was a really relaxing weekend in which we stayed in log cabins and water skied on the Ozark Lake.

The break was just what I needed to get me ready for the jam packed final two weeks of basic training. I had to pass the basic training test, confidence course, road march, and last but not least our final three days in the field known as "Bivouac."

The final basic training test was a series of stations consisting of first aid, general orders, saluting, addressing an officer, weapons, and so on. At each station you would be tested and had to pass that station in order to proceed. For example, in the case of the general orders station I had to enter and state my general orders, in the case of the weapons station I had to tear apart the M16 rifle and reassemble it in a certain amount of time. Of the 190 left in the company only 50 passed it the first time, me being one of them. There was one more chance and all but four passed. The four failures were reverted back to another company.

The confidence course was a series of huge wooden obstacles that had to be overcome one by one. Some were small and others huge structures that got harder and harder as you went on. As you climbed up and down these barriers you would progressively get worn out, making the next one even that much harder. Some of these were low to the ground but others towered 30 feet in the air, and you had to have confidence in yourself to get through it.

You're always told to never leave a comrade behind, and that seemed to be the point of the road march. We loaded up with all of our equipment in our rucksacks along with our helmet, gas mask, and weapon for the 20 or so mile march. It was fast paced, almost like run walking, which caught a lot of us off guard. As we went along in two columns spread out along the road, it wasn't long before there were injuries such as pulled muscles. These guys were loaded up into the medical vehicles that trailed the march. Those guys who were simply fatigued started to fall behind and were quickly picked off by the drill sergeants who were stalking behind the formation like a pack of malevolent wolves.

As guys started to fall back, we started to take some of their gear off of them to make it easier for them to keep up. No one wanted to fall back and into the jaws of the drill sergeants. That would assure being reverted. Toward the end of the march we were more of a collection of guys carrying and helping each other along rather than a structured formation on a road march. Although there were losses, we made it pretty much as a group. That appeared to be the goal as the drill sergeants seemed content.

Bivouac was the last event in boot camp. It was three days out in the woods pretending like we were at war. The weather was getting pretty mild in late September so being outside wasn't so bad. Our days were spent pitching tents and pretending we were hiding in the woods from the enemy. Our nights were spent standing guard and protecting our weapons from being taken by a drill sergeant. It was drilled into us that "you will always have control over your weapon!" If you were to lose it, you were done. And the drill sergeants were always on the prowl for an unsecure weapon.

One night there was a ruckus and it turns out one of the drill sergeants had tried to sneak into one of the two-man tents and take a weapon from a sleeping recruit. The recruit woke up and kicked the drill sergeant in the face! Nothing was done over the incident, just a scared recruit and an embarrassed drill sergeant!

On Friday the 17th of September we finished boot camp and would start advanced individual training (AIT) on Monday after a pass over the weekend. I spent those days on pass lying around watching TV and drinking beer in a Rolla, Missouri motel room with four of my buddies. Rolla was about 40 minutes away and the small and dated motel seemed like paradise away from the rigors of basic training.

Monday the 20th of September was our first day of AIT and I remember the day very vividly. We were all standing out in formation on the concrete pad in front of our barracks for the usual 15-20 minutes waiting for the drill sergeants to come out and direct us in our daily routine. The famous SSG Strickland came out and stood on the elevated landing that led up to his office, lit his trademark long cigarette, and said in his unique robotic drawl, "You are no longer in basic training, you are in AIT." He then turned and walked back inside.

Learning how to tie knots in AIT

Our routine stayed pretty much the same throughout our basic and AIT training, just the type of training changed. I really would have never known the difference if not for SSG Strickland's proclamation. Instead of non-stop training on shooting, marching, and basic soldiering, we moved on to bridge building, knot tying, explosives, and land mines.

As we neared graduation from AIT some things like going to the PX were lightened up and we were allowed to go by ourselves instead of marching as a company or platoon. I really didn't like this and preferred transiting with my company and drill sergeant. If we passed a drill sergeant or an officer as a company, the drill sergeant leading us would handle all the formalities like saluting. When out on my own I had to handle the formality and that scared me, especially if I ran into an officer! Would I salute correctly and at the correct time or get my ass chewed. What if I ran into a drill sergeant! Would I be able to stop fast enough and come to the position of parade rest and let him pass?

It was like running another gauntlet whenever I had to go someplace. It didn't take long for my fear to materialize. I was standing outside of the big main PX with a friend and an officer walked out toward us. I got scared and turned away like I had not seen him. It must have been pretty obvious because he walked up to me and chewed my ass for turning my back and not saluting or addressing him.

All of us in boot camp and AIT were in the military occupational specialty (MOS) 12 series and had to learn the basics of that series.

After graduation about 90% of the company would go on to be MOS 12Bs, Combat Engineer, which specialize in handling the explosives and land mine duties. I was a MOS 12C, Combat Engineer Bridge Crewman, and would go on to specialize in building combat bridges. There were actually two other guys in my company, who were also MOS 12Cs, and would transfer on to Korea with me, Mike Guertin and Terry Baer.

On the 28th of October we had our graduation ceremony and officially completed AIT. I was now a combat bridge crewman MOS 12C, making $541.40 a month, and a private in the United States Army. All I had to do now was out-process and await my travel orders. Part of out-processing was turning in my non-personal gear that was assigned to me such as my tent half, rucksack, canteen, helmet, locker, mattress and other things I had no memory of ever having signed for. I guess on the first day I arrived I had signed for my mattress and locker. Now if you remember how I detailed the first day of boot camp you will understand I would have signed anything for any reason.

Well, the drill sergeant came to inspect my mattress and locker to make sure it was in the same condition in which I had received it. He flipped my mattress over and there was a stain on it about the size of a small pizza. It looked like someone had either soiled it or spilt metal cleaner on it. The drill sergeant in an annoyed tone of voice asked something to the effect of, "How did this get here Semler?" I'm thinking; how the heck do I know, I didn't put it there! Of course, I don't even remember signing for a mattress in the fog of the first day and I couldn't see myself having any reason or desire to flip it over to inspect it. The drill sergeant was not happy and told me I needed to carry that mattress down to the supply room, which was about a mile away, and show them the mattress.

So, I cart this twin-sized mattress all the way down to the supply room and explained to the supply clerk that the drill sergeant had sent me down to show them the mattress. They frankly told me to get my ass and my mattress out of their supply room! So, I carted that mattress all the way back up to the barracks and up three flights of stairs to my awaiting drill sergeant, who proclaimed that I had just bought a mattress!

I never did see it come out of my pay so I'm not sure how I ever paid for it except in embarrassment carrying it around the base.

There was a lull in between graduation and the date I was to ship out. Most of us being held longer were deploying overseas and needed more time to out-process. These days were filled with watching movies on venereal disease, shots, and lectures on deploying overseas. The atmosphere was relaxed as the drill sergeants were taking their leave and standing down in between basic training companies. They were still ever present, and I avoided them at all costs as before. I took this opportunity to sneak down to the 1st floor area and have my picture taken with the "Bravo Bull." The mural was located right next door to the drill sergeant's main office, which made me extremely nervous. This area was totally off limits during basic training & AIT. And I didn't even know if I was allowed to have a camera.

So, I scurried down to the first floor with a willing friend to have my picture taken. I was a bundle of nerves. We quickly walked down the hall, posed, and got the hell out of there!

Me with the Bravo Bull

After out-processing and receiving my travel orders I boarded a Greyhound bus and departed Fort Leonard Wood on the 3rd of November. After taking 15 days of leave at home in Pennsylvania I departed for Korea on the 17th of November 1982.

CAMP PELHAM KOREA
1st PLATOON, E COMPANY, 2ND ENGINEER BATTALION,
2nd INFANTRY DIVISION
18 November 1982 – 18 November 1983

I arrived in Osan, Korea on the 18th of November 1982 after a long 18-hour commercial flight from Pittsburgh, Pennsylvania. The flight made stops in St. Louis, Los Angeles, and Tokyo, Japan for refueling. The sight of snowcapped Mt. Fuji as we circled for our landing in Japan is still vivid to this day. After arriving in Korea, I was taken to the 2nd Infantry Division in-processing center at Camp Casey in the city of Tongduchon.

The processing center was known as the "Turtle Farm." They called it that because the in-processing and out-processing buildings were right next door to each other, and it took your year-long tour to get from one building to the other. Hence, you were referred to as a "turtle." This nickname would stick with you for about the next six months. It was also slang for doing something stupid. The phrase, "You dumb fucking turtle," was a commonly used good example. After your first six months you morphed from a "turtle" to a "short timer," known as just plain being "short." Unlike "turtle," guys were very proud of this designation and would walk around holding up their thumb and index finger almost touching saying, "short! "The closer your fingers where to touching, without actually touching, the shorter you were.

The Turtle Farm was a relaxing place in itself and I don't remember going off-base, or if I was even allowed to venture off into the economy. We did spend our free time going to the bowling alley on base and that's where I first encountered human pin setters. As you rolled your ball down the alley there were these young Korean kids who sat up above the pins out of sight. After your ball hit the pins, these kids would hop down and reset your pins and roll the ball back up the alley to you. At the end of bowling, we would throw money down the alley as their tip.

Most of our day was spent in-processing, which entailed lectures on the country, shots, and everything else needed for our new unit before we were shipped out. During one of these lectures they discussed the two Army divisions stationed in Korea, the 2nd Infantry Division and the 8th Army. I was so amazed at how cool the 2nd Infantry patch looked over the plain 8th Army patch. I wasn't really sure where I was at, and hoped I was heading to the 2nd Infantry Division, because they had a better-looking patch. An obvious priority of an 18-year-old! As luck would have it, and since I was already at the 2nd

Infantry Division in-processing center, that's where I was headed.

The departure day arrived, and I loaded up into a two and half ton truck, known as a "duce-and-a-half" with a bunch of other turtles. The back of the truck had a long wooden bench seat on each side with a canvas cover. This was in November and there was no heat in the back, so we were freezing. The weather and seasons in Korea were very similar to that of Pennsylvania. The truck, loaded up with turtles and our belongings, headed out the gate of Camp Casey on its way to various destinations within the 2nd Infantry division's area of operation.

The 2nd Infantry Division operated in the western corridor of South Korea around the Demilitarized Zone (DMZ). The DMZ is what separates South and North Korea and is a heavily fortified area. The DMZ cuts from one side of the country to the other. The United States 2nd Infantry Division guards the western half of the DMZ and the South Korean Army covers the eastern section. The section of the DMZ guarded by the 2nd Infantry Division is considered the most strategic. This is because Seoul, which is the capital of South Korea, is located in the western section.

The 2nd Infantry Division acts as a buffer between North Korea and strategically important Seoul.

This was a combat infantry division and no women were assigned to the units up on the DMZ. The area was also considered a hardship tour, and you were not authorized to bring any family members. The Army strongly discouraged the practice. But some of the guys, mostly NCO's and officers, who were married to Koreans, had their wives and families meet them and settled near the camp. There was no family support network, so if you brought a spouse and family, you were on your own finding a place for them to live. And if the shit hit the fan, they were on their own.

View of the DMZ and North Korean guard post

Since the cease fire in 1953, 29 years prior to my arrival, there have been numerous incidents and deaths of United States and Korean soldiers who patrol this area. The DMZ, when I arrived, was pretty much where it was when the ceasefire was established in 1953. There was no peace agreement with North Korea, just a ceasefire. So hostilities could start up at any time for any reason.

This area is also known as the 38th parallel because of the DMZ's proximity to it. Toward the end of the war, when both sides seemed to be in a stalemate around the 38th parallel, some of the fiercest fighting occurred at places such as Pork Chop Hill, Bloody Ridge, Old Baldy, The Punchbowl, Heart Break Ridge, and Outpost Harry. The area Camp Pelham occupied had been held by North Korean and Chinese forces several times as the war waged back and forth. The battles previously mentioned straddled the current DMZ and some occurred on the South side and some on the North side.

The 2nd Infantry Division had about 20 camps strewn out amongst the countryside north of Seoul up to the DMZ. These little camps had two or more combat companies assigned to them. It was my understanding that wherever a company was dug in during the cease fire was where they pretty much remained. I was headed to a place five miles from the DMZ by the name of Camp Pelham. Camp Pelham was about a mile and a half east of a major town named Munsan in a little town called Sunyu Ri. The Korean spelling was Seonyuri but for some reason it was spelled Sunyu Ri by U.S. forces. I would soon find out my movement in this area would be regulated by passes and curfews, and I would be held to rationing protocols.

As the deuce-and-a-half rumbled out of Camp Casey we turtles were huddled in the back trying to keep warm from the freezing temperatures outside. Every now and then we would pull back the tarp covering the back and see where we were. The countryside was made up of semi-paved roads, dirt roads and small shacks. The little towns we passed were just a grouping of more dirt roads and shacks.

As we rumbled on, we lost more and more guys as the driver made stops at various outposts on the way to Camp Pelham. It began to get dark and a peek out the back flap revealed total darkness and I thought to myself, where in the heck am I going? Finally, at the very end of his route, our driver pulled into Camp Pelham and dropped me off along with the only other passenger still on the truck, at what would be our new home for the next year.

The Camp was named in honor of John C. Pelham, a young Confederate artillery officer who served under J.E.B. Stuart during the Civil War. He was very popular and known as the "Gallant" Pelham for his bravery. Unfortunately, he was mortally wounded at Kelly's Ford, Virginia in 1863, dying the next day at the age of 24.

Camp Pelham Korea 1983

The camp was situated in the middle of a huge rice patty with a bridge connecting it to the little road that ran through the little village of Sunyu Ri. There wasn't a lot to Sunyu Ri, just a lot of low-level structures. Vehicles were limited in this farming community and most of the time when you saw a car it was a taxi. The locals all drove a sort of rototiller on two wheels that were connected to a cart. The man would drive, and the family would be in the cart.

Downtown Sunyu Ri with the typical rototiller vehicle

Sunyu Ri was a very friendly town and community. I never felt fearful of the civilian population there and they always seemed to be appreciative of our presence. We provided an economy for them with jobs at the camp and patronage of the village bars. And any time we could offer humanitarian assistance, we did.

Like on the 21st of August 1983 when a local village boy was hunting the poisonous Salmosa viper valued for its health and medicinal value. The boy was hoping to cash in on the 30,000 won, $50.00 USD, the snake would bring. He was unfortunately bitten on his hand. Because the camp was the closest medical facility he was taken there for treatment. When he arrived the U.S. medical personnel realized he was in bad shape and he was medevac'd by helicopter to the 121st Evacuation Hospital at Yongsan, near Seoul. The fast action by the Camp Pelham aid station saved his hand and maybe his life.

And when I arrived right before Thanksgiving, one of the first things I did at Camp Pelham was sponsor an orphan to come on the camp for Thanksgiving dinner. It was another great way to give back to the local Koreans.

Me & my orphan on the far right

The buildings on camp were very basic, single-story structures, spread out over the compound which held Echo Company, 2nd Engineers and Bravo Battery, 2/17th Field Artillery. I was assigned to the "River Rats" of the 2nd Engineers which consisted of four officers and 117 enlisted men. The 2/17th field artillery had similar manning and maintained a fire base known as 4-Papa-1, which was the only active firebase in the Army at the time. Their mission was to shell invading North Korean forces with their 105mm and 155mm howitzers.

Living conditions were very basic. All the barracks were one story structures of simple wood and metal construction and heated by oil furnaces. We had open bay style living with metal bunk beds and metal stand up lockers, which acted as walls between each set of bunk beds. The showers, toilets and sinks were in an adjacent building known as the latrine. The latrine was always interesting to get to in cold weather, which seemed to be year-round. The normal mode of operation was to get naked with your towel wrapped around you, shower flip flops on, and bath kit in hand. You would brace yourself for the cold, then swing open the barracks door and bolt out the door as fast as you could to the latrine. Once there you would find a row of sinks, row of toilets, and a huge open bay shower. Check your modesty at the door!

South Korea has mandatory service which requires all males to serve in the military. Those that were highly educated and attending college were the most likely to become Korean Augmentation to the United States Army (KATUSAs). This was because only those that score very high on the Korean Service exam are assigned as KATUSAs. KATUSAs were assigned to U.S. military units to assist in translation and help in the day-to-day activities of getting around the country. Like every unit we had KATUSAs all throughout our company, with several assigned to our platoon and one to our squad. The

KATUSAs worked alongside us, slept in our barracks, ate in the chow hall and shared in most all of the privileges we had, with the exception of PX goods. Although they did the exact same work as us, a KATUSA sergeant only made 4,500 won, or $5.90 USD a month, compared to $760.80 USD a month made by a U.S. military sergeant!

The KATUSA assigned to my squad, and who I shared a bunk bed with, was Joo Myung Ho. He was a corporal for most of my tour and was promoted to sergeant before I left. Like all the KATUSAs he had a great sense of humor and always maintained the highest standards of Korean honor and tradition.

Corporal Joo Myung Hoo

Their English was good enough to get them by and our language barrier was the only thing that kept us from really long and detailed conversations. For the most part the KATUSAs didn't share a lot of personal information and kept to a more professional relationship. They didn't go into the village with us drinking or whoring around. They had their own little building on the camp where they met and drank. We would always say they reeked of kimchee and Jinro, a favorite food and liquor, when they came back at night from their hangout.

Those of a lower education, and usually of a lower standing in Korean society, are entered into the Republic of Korea (ROK) forces. The ROK service was very tough and there was no tolerance for disobedience.

I was riding in a bus from one camp to another when we got stuck behind a group of ROK soldiers road-marching down this road with all their gear on. One of the soldiers was trailing behind and he was followed by a senior soldier beating him in the back with his weapon every time the straggler tried to slow down. I don't mean beating him lightly; I mean he was beating him to hurt him. This went on for quite a while until they turned off the road. I can't imagine what happened when they were out of sight.

On another occasion while I was standing guard duty on a drill bit stuck in the ground (more on that later), I walked over to watch the neighboring ROK soldiers training with mortars. They were training on how fast they could put the mortar together and then disassemble it. Whenever they messed up, they had to take a piece of this very heavy mortar and do pushups with it. After watching for a few minutes, I had to walk away because I felt sorry for the poor guys.

The camp had primarily semi-paved roads and a few sidewalks. When it rained or snowed everything turned to mud. On the edge of the roads on both sides were these two feet deep by two-foot wide concrete square ditches that lined the roads and were for catching water run-off. They were actually quite treacherous if you miss-stepped. We called them, "turtle traps" because the new guys always seemed to take a tumble into these at least once.

There was a small chapel, mess hall, club, barber shop, tailor shop, and associated military buildings such as armory, and orderly room which had the only TV around. The TV was black and white and only brought in the single Armed Forces Network (AFN) channel. The orderly room was the hub of all activity for the company. It's where we picked up and turned in our passes when leaving the base and hung around on the cherished sofas waiting for friends to get changed to go out on pass. It's also where we got paid every month and collected our mail. The company clerk and staff were located there and right next to the orderly room was the armory, where we checked in and out our weapon and gas mask.

On the camp there was also a photo shop ran by a Korean guy named "Snap." Snap would patrol the camp, village, and motor pool, taking pictures of anything, or anyone, he thought he could get money out of. Once a week you would swing by and see if Snap had any pictures of you and purchase them. He also made custom albums made of laminated wood and mother of pearl inlay. I bought one customized for me before I left.

There was an annex on the other end of the village named RC4, which housed the PX and several other U.S. military units. Because of the extensive black marketing of American goods everything American made was rationed. I was issued a ration card and was limited on the amount of cigarettes, alcohol, and just about everything else I could purchase in a month. This was serious business to us because we all smoked like freight trains and drank like fish.

Dallas Cowboy Cheerleaders

We had a nice club on the camp with Korean employees. It was mostly used as a backup watering hole if you couldn't go into the village because you couldn't get a pass or were on restriction. It wasn't that bad as clubs go and they always had something going on; they even had a pole dancing stripper as entertainment! It was also a gathering spot around the holidays as they always had a tree up and the place decorated for the season. Soon after I arrived the Dallas Cowboy Cheerleaders visited the club on a Christmas U.S.O. tour.

What a surprise that was. I was detailed to put up a huge, 32 feet long by 16 feet wide, tent known as a GP Medium to accommodate the overflow of food and drink for the event. As you can imagine they drew a huge crowd!

There were also reminders that we were close to the DMZ, like the bomb shelters next to the barracks. They were made of concrete and covered in grass. We had a picnic table on top of ours and used it as a nice place to drink beer and gaze over Camp Pelham and Sunyu Ri.

Bomb Shelter to the right

And when winter settled in and the ground froze solid, you would occasionally hear a faint blast. I was told the sound was a landmine exploding at the DMZ. The weight of the frozen ground would set them off.

Our sole mission there at Camp Pelham was to build an escape bridge over the Imjin River, which was about a mile away. The purpose of the bridge was to evacuate civilian personnel on the other side of the river and military personnel in the Joint Security Area at Panmunjom, Camp Greaves, and Camp Liberty Bell.

Panmunjom is where talks between South & North Korea were held and Camp Greaves & Liberty Bell were infantry units who patrolled the DMZ were stationed.

There were only two permanent bridges going over the Imjin River and they were both wired for demolition should an attack from the North occur. The United States Army controlled Freedom Bridge and the ROK Army controlled Liberty Bridge. Our sister engineering company at Camp Edwards East, a few miles away, maintained the explosives on Freedom Bridge that would be used to blow it if needed. The ROK Army maintained the explosives at Liberty Bridge. With these two bridges blown up it would be up to us to erect our pontoon bridge and evacuate anyone left on the other side. The reason for destroying the fixed bridges was to prevent North Korea from crossing them, hence slowing their advance.

Bridge constructed across the Imjin River

I had the opportunity to go over Freedom Bridge once. Our deuce-and-a-half driver had broken his arm, and they needed a replacement driver to shuttle some guys from another unit over to the live fire range on the other side of the Imjin River up around the DMZ. This sounded like something different, so I volunteered.

The task was to take these guys over Freedom Bridge to the firing ranges, let them do their thing, and bring them back. Going over Freedom Bridge was exciting in itself. As I mentioned earlier, the bridge is kept wired for demolition 24/7 by our sister engineering company of MOS 12Bs. There is nothing like the feeling of driving over a bridge wired with tons of explosives!

We loaded up and headed out on a nice sunny day and made our way over Freedom Bridge and down along the DMZ.

Freedom Bridge

The first stop was at the M72 LAW rocket launcher range. These guys shot a bunch of these LAW rockets at a tank downrange while I took a cat nap in the truck. When they were about done, they called me over and asked if I wanted to shoot one. I had never fired a real one before and said "Sure." It was a really cool experience to have that rocket shoot out of the shoulder fired launcher and wiz down range to the awaiting tank and explode. Nothing like the bottle rocket in boot camp!

The next stop was the M203 grenade launcher range. This is mounted to the underside of the M16 and fires a grenade. Now that my adrenaline was pumping from the LAW rocket launcher experience, I asked if I could fire a few grenades down range. It was another great experience. And these weren't filled with paint like boot camp; they were the real thing and exploded!

On our way back we got close to the DMZ and I could see the barbed wire fencing and observation towers going on for as far as the eye could see. This was it, the actual DMZ. It gave me an eerie feeling knowing North Korea was just on the other side and could take a shot at me or anyone else at any moment. It was the only time I ventured across the river and had no desire to go back.

Alert sirens were sounded whenever there was a threat of North Korean aggression. We were right up on the DMZ, so alerts were always taken very seriously. When an alert was sounded you would hear the alert siren wailing away and we would go into battle mode. The whole area went into an immediate lockdown, and we would scramble to get our weapon, gas mask, head to the motor pool, and roll out to the Imjin River.

These could be anything from infiltrators to a plane flying into South Korean airspace.

This happened a lot, and each time we never knew what was going on. But the urgency to get to the motor pool and get rolling to our combat position, the Imjin River, was high. Every alert was an adrenaline rush, and every unit was rushing to their combat position. The 2/17th called them "speed balls" and rushed to 4-Papa-1, our sister engineering company of 12Bs rushed to Freedom Bridge, and so on. There wasn't a shortage of adrenaline rushes in 1983.

X – Is the location of Camp Pelham

On Friday the 25th of February we went on alert when there was a report of a North Korean fighter jet flying over the DMZ and headed for the capital of Seoul. It turns out it was a defection and not an attack. Twenty-Eight-year-old North Korean Air Force Captain Lee Ung-Pyong was assigned to the 1st North Korean Combat Air Division flew his Chinese built MIG-19 fighter jet across the Yellow Sea.

Capt. Lee Ung-pyong

He said he took off at 10:30 am from Kaechon Air Base, which is about 160 miles north of Seoul. He was flying with another MIG-19 fighter on a rocket firing practice mission. A few minutes after taking off he said that he broke away from his partner and sped south at about 3,000 feet going 531 mph. As he raced south, he was hailed via radio calling him to turn back. This caused him to dive to only a few hundred feet above the ground maintaining an air speed of around 575 mph. When he crossed the DMZ at around 10:45, near Haeju, he was 80 miles northwest of Seoul. He was then intercepted by ROK fighter jets who escorted him to the ground. This was only the 5th such defection since 1950.

He said he defected to warn South Korea that the North Korean Communists were pushing war preparations in a frenzied manner, claiming war was the only way to unify the divided Korea. He said things escalated as the South started preparing for the huge military exercise known as Team Spirit 1983.

On May 5th the highly unusual hijacking of a Chinese jetliner with 105 people on board placed us again on high alert. The civilian plane was hijacked by six Chinese defectors while on a flight inside China from Shenyang to Shanghai. The hijackers stormed the plane's cabin and shot and wounded two crewmembers and forced the captain to divert and fly over North Korean airspace in route to Taiwan. The pilot tried to trick the hijackers and land at Pyongyang, the capital of North Korea, but the hijackers caught on and derailed the attempt.

As they flew over the Korean DMZ the plane was intercepted by South Korean fighter jets and escorted to the U.S. military base at Camp Page Korea near the mountain resort city of Chuncheon. This would be about 40 miles west of Camp Pelham and 45 miles northeast of Seoul. The Chinese immediately demanded their plane and people be returned. This was a little problematic since there are no political ties between China and South Korea since they fought each other during the Korean War. The Chinese were currently supporting the North Koreans as they did during the war.

Taiwan was urging the South Korean government to retain the hijackers as they would be immediately executed if returned to China, which had happened to the last hijackers of a Chinese plane. The hijackers were eventually tried in South Korea, receiving sentences ranging from four to six years and the 99 passengers and crew returned to China.

A North Korean Army captain, Captain Shin Chung Chol, walked across the DMZ near Yanggu about 60 miles northeast of Seoul on the 7th of May sending us on alert. A South Korean Army patrol spotted him and he indicated he wanted to defect, and was led to a nearby guard post. The captain was from the 13th North Korean Army Division and said that life was hard in North Korea. Later on that year on the 1st of August Captain Shin was commissioned as a major in the South Korean Army.

On the 19th of June we were once again on alert. Just a mile away from Camp Pelham, in Munsan, three North Korean infiltrators were killed crossing the Imjin River which we bridged. They were carrying Russian made pistols, Czech made submachine guns, sophisticated silencer pistols, daggers, knapsacks, maps, radios, binoculars, 500,000 won ($680.00 USD), and South Korean Army uniforms. In a hail of rifle fire and grenades they were killed by ROK soldiers.

In August three incidents occurred within eight days of each other placing us on alert. The first involved the killing of four North Korean frogmen on August 5th.

Four dead North Korean frogmen

The North Koreans had disguised their 60-ton vessel to look like a Japanese fishing boat complete with a Japanese name. A South Korean patrol boat became suspicious and when it engaged the vessel a fire fight broke out and the North Korean vessel was sunk. It was thought that this vessel was a mother spy boat for other infiltrators because of the equipment on board and a secret gate type device on the stern to launch and recover smaller boats. Along with the bodies, they discovered two Japanese radars, one 82mm recoilless gun, and two 14.5mm anti-aircraft machine guns.

In a more dramatic defection on the 9th of August a Chinese Air Force colonel flying a MIG-21 crossed the Yellow Sea. He was seeking asylum in Taiwan which had a standing reward of 9,310 ounces of gold, worth about 3.85 million dollars at the time, for any Chinese pilot who defects with a MIG-21. It was originally thought that the plane was part of a North Korean invasion and South Korea issued the first air attack warning over Seoul since the cease fire with the North in 1953. It ordered all citizens of Seoul and surrounding areas to take shelter and turn off all electricity. Obviously, this caused wide spread panic as all cars on the streets were stopped, sending passengers fleeing. It was only the third Chinese Air Force defection since 1953.

The pilot Colonel Sun Tien-ching later said that he defected because his father was tortured to death while in prison in 1968. He said he left behind his mother, wife, son and daughter. The pilot, who earned $60 a month in China, was given a promotion to colonel in the Taiwan Air Force and received his reward of 9,310 ounces of gold, making him an instant millionaire. South Korea ended up keeping the Mig-21 but Taiwan decided to give him the money for his heroic defection.

The third incident involved the killing of at least three North Korean boat crewmen on a spy boat on the 13th of August. Like on the 5th of August this vessel was disguised as a Japanese vessel and refused orders to heave to and began to flee at 40 knots from South Korean forces. Three intact bodies were recovered, parts of bodies, communist weapons including machine guns, dive suits, and notebooks with pictures of the North Korean leader Kim IL Sung and his son.

On the 2nd of September Korean Air Lines (KAL) flight 007, a Boeing 747, carrying 265 passengers from New York to Seoul was missing and reportedly intercepted by Soviet fighters and forced to land. One of the 61 American passengers was United

States Congressman Larry P. McDonald, a democrat from Georgia. He was heading to the 30th anniversary ceremony of the signing of a defense pact between South Korea and the United States. He had missed an earlier flight with several other congressmen and ended up on KAL 007. The next day the KAL plane was found to have been shot down by a Soviet SU-15 fighter jet as the 747 strayed over the Soviet island of Sakhalin, killing all 265 on board. It was the fifth worst aviation disaster in history.

South Korea did not have diplomatic relations with the Soviets. Since the plane was shot down near Japan, and there were numerous Japanese nationals on board, the Soviet ambassador to Japan was immediately called in to explain what happened. He explained that the plane had crashed and reports of it being shot down were propaganda. The Japanese however had overheard radio transmissions between two Soviet fighter jets that intercepted flight 007 and one had radioed to ground control that he had fired a missile, and the target was destroyed. It was determined that one of the Soviet fighter jets was trailing behind the KAL airliner. It fired one heat seeking Anab rocket, which brought the 747 down.

The Soviets would later admit that they mistook KAL 007 for an American RC-135 reconnaissance aircraft. The RC-135 was operating within 75 miles of the 747 when the passenger jet mistakenly crossed into Soviet airspace.

The whole world condemned the attack and was outraged. The unprovoked downing of flight 007 hit hard with us at Camp Pelham as we had all arrived in Korea via commercial planes. Possibly on KAL flight 007. It scared me because I was getting short and would be heading home in a few months via a commercial carrier.

This downing was eerily similar to KAL flight 707 a few years earlier. It was traveling from Paris to Seoul and carrying 110 passengers on April 20th 1978 when it was shot down by a Soviet Mig fighter jet. It luckily was able to crash land on a frozen lake in Murmansk, a port city in the USSR, killing only two passengers. The Soviets claimed the KAL jet had strayed over its airspace and was shot down after refusing to acknowledge the Mig fighter's demands to land.

With both KAL flight 007 and 707 the airliners had strayed off course and inadvertently were flying over very sensitive Soviet military installations. The Soviets made accusations that the planes were spying, and this is thought to be the reason why they were shot down. South Korea did not have diplomatic relations with the Soviets and relied on other nations such as Japan and the United States to step in and to negotiate on their behalf.

On the 9th of October all of South Korea was placed on high alert when the President of South Korea, Chun Doo-hwan, narrowly escaped an assassination attempt while visiting Burma. The blast killed 21 people including 17 South Korean dignitaries while 48 people were wounded. Four of the 17 South Korean dignitaries killed were cabinet members. The only reason President Chun was unhurt was the fact that his motorcade was stuck in traffic and arrived five minutes late to the event. The three captured suspects were later identified as North Korean commandos sent to assassinate the South Korean President.

Leaflet showing North Korean assassins attacking South Korean President Chun

I found the above leaflet around Camp Pelham, delivered by a North Korean balloon. On the right it says, "The ghost that follows you" and the words on the two North Korean assassins say "Anti-Chun Doo-hwan power." The scared lady and gentleman in the center are South Korean President Chun Doo-hwan and his wife.

The tension on the DMZ was very high after the assassination attempt and the North Koreans fired over 2,000 rounds from two of their guard posts on the 14th of October. The North had claimed that 10 South Korean soldiers crossed into the North firing 500 rounds. President Chun accused the North of plotting to invade the South.

Luckily no one was hurt in the one sided firefight. I know the guys up at outpost Ouellette and Collier, the two northern most American guard post on the DMZ, must have thought WWIII had started!

Every now and then we would wake up to find North Korean propaganda spread all over the ground at either the barracks or motor pool. The North Koreans would send balloons filled with the small leaflets over the DMZ into South Korea that would burst at some predetermined time in flight. The KATUSAs assigned to us would scurry around and scoop them all up and tell us not to look at them, of course that made us want to see them!

This is a typical one that I found one morning down in the motor pool. It shows a South Korean ROK soldier who defected to the North on the 11th of September 1983.

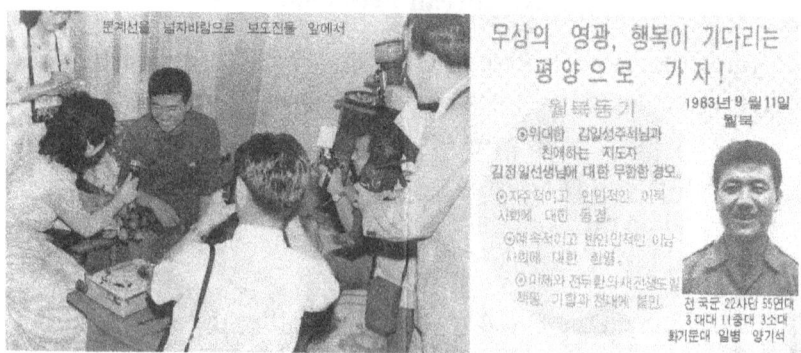

North Korean propaganda leaflet

On the right it says, "The highest supreme honor, let's go to Pyongyang, where happiness is waiting!" In the center it reads "Motive for Defection: Boundless joy and admiration for the glorious President Kim Il-sung and for the beloved leader Kim

Jong-il. He fantasized about the Socialist life to the North and became disillusioned with the anti-citizen ways of the South. He is against the Americans and the war-mongering regime of Chun Doo-hwan, who actively seeks to restart a war."

I'm sure that's what they want us to think. And by the looks of their not- so-state-of-the-art camera's and recording equipment, even for 1983 standards, I'm sure things were not all that great in North Korea!

The defection that was talked about the most in my circle was not from the North, but from the South, and involved an American. PFC Joseph T. White, with the 2nd Infantry Division, shot the gate door padlock off with his M-16 rifle at guard post Ouellette in the Joint Security Area of the DMZ on the 28th of August 1982. Although he made it into North Korea, he left behind a duffle bag with a camera containing undeveloped sensitive film of the bunker and tower system of his post and information about American radar sites in the area.

He was apparently distraught about not being granted leave to visit his sick South Korean girlfriend who was hospitalized in South Korea. The 21-year-old was reported to have brushed up on his Korean language skills before defecting. His defection was the first by an American in 17 years and the fifth since the Korean War fighting stopped in 1953. In 1985 his parents received a letter from a North Korean friend of PFC White's stating that he had drowned in August of that year in the Ch'ongch'on River in North Korea. They requested his body, but it was never returned to White's parents and was reported to never have been recovered from the river.

There was also the stress of being up on the DMZ that just drove guys nuts. In June of 1983 Private Mark A. Burford, who was assigned a few miles away at Camp Greaves, checked out his M-16 rifle along with ammunition and took several hostages. Like PFC White it seems he was distraught about not being able to see his South Korean girlfriend. After an eight-hour standoff he surrendered to authorities at the camp. PVT Burford had only been in country six months, since the 6th of December 1982. Heck he was probably only a week behind me at the Turtle Farm.

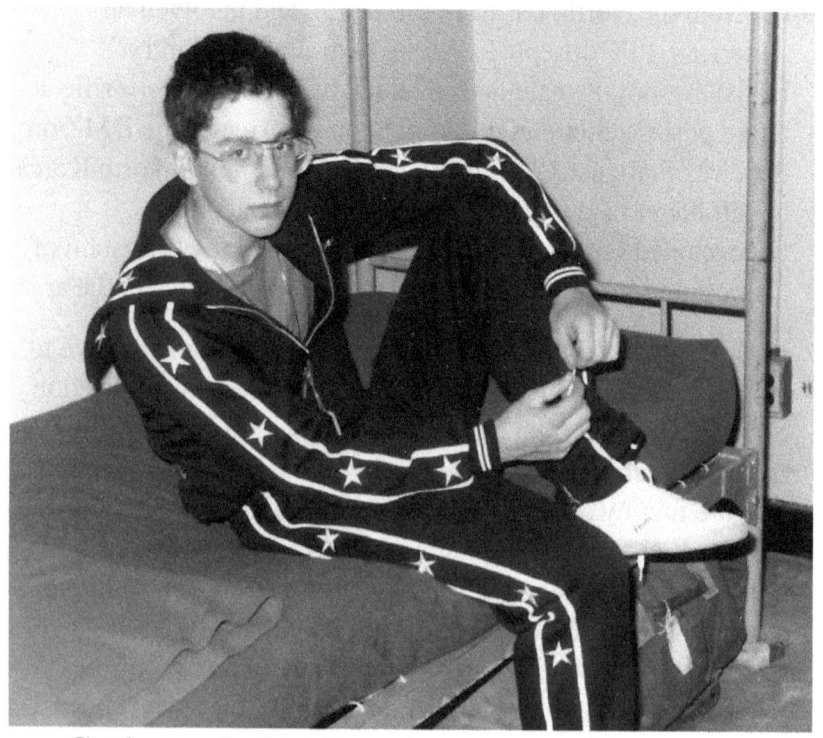

Getting ready for PT in my custom-made PT outfit

Our daily routine while we were at the camp was up at 0500 for physical training (PT) and a two mile or so run out into the village and surrounding countryside. It didn't matter what the weather conditions were, we ran. This was always a colorful event. The camp was too small to run in so we would always run out the gate, through the town, and into the countryside. We didn't wear our military uniforms but instead we wore tailored PT outfits made by the local Korean tailor. Everyone had their own style and color. Think 80's disco!

And when you run in the Army you sing cadence like in boot camp to keep everyone in step. We sang your typical military authorized cadence like, "Airborne Ranger" which was one of my favorites. It went sort of like this:

C-130 rolling down the strip,
Airborne Ranger going to take a little trip.

Hop up, buckle up, shuffle to the door,
jump right out on the count of four.

If my chute don't open wide,
I have another one by my side.

If that one don't open either,
I have a date with ole Saint Peter.

If I die in a combat zone,
box me up and ship me home!

Along with a nice tune like that we mixed in local raunchy cadences about the prostitutes and bars that were just vulgar. I guess it really didn't matter because the locals didn't understand English.

When I first arrived at Camp Pelham, I was put in a break-in PT detail. We would run in the evenings so that we would build up our running endurance and be able to hang with the company's longer runs in the morning. The break-in running wasn't bad but what really worried me was running out along the country roads in a small group of four or five. I was new to Korea, and it just seemed odd to be running along these dirt roads between these huge hills so close to the DMZ. Especially in neon sweat suits! With all the North Korean incursions in the area I felt we were sitting ducks for them to take us out.

Once back from a run the first thing I did was smoke a cigarette. It sounds weird but that smoke after a long run tasted so good and gave me a slight burning sensation that was actually pleasant. Along with my smoke I would mix up a glass of TANG orange drink and gulp it down.

After our morning PT run, my cigarette & TANG, it was a shower, breakfast, and then formation down at the motor pool for work.

Meals at the mess hall were about our only daily interaction with the 2/17th field artillery. We shared a common mess hall and that was about the only time you wanted to engage them. The threat was from North Korea but the 2nd engineer's nemesis' was the 2/17th. I'll get more into that later.

Then it was down to the motor pool, which was just outside the camp to the right at a little annex. Once there we would inspect our trucks, start them up, and then sit in them while smoking and telling stories about back home. Guys that didn't have a truck assigned to them would hang out with someone that did. At some point the snack truck from RC4 would come rolling through and we would all pile out of our trucks and get a snack. After our snack we performed maintenance on our truck and bridge equipment.

My truck, E-109 "Best on Line"

When I was assigned my own truck, a few months after arriving, it didn't run, and it didn't have a roof on the cab. It was truck number E-109 and everyone referred to it as "109 dead on line." I referred to it as "109 best on line," sort of as a joke.

After a few weeks E-109 finally made it into the motor pool garage to be worked on; and it spent a lot of time in the motor pool garage getting worked on! The first time it made it out of the garage it only made it a few feet. The mechanics had parked it just outside the garage facing the huge 30 feet high by 20 feet wide garage doors.

That night the alert siren went off. Now when the alert siren went off there were always a few guys detailed to run to the motor pool and start all the vehicles while the rest of us drew our equipment. One of the guys on this detail was a great guy from Boston named John Sodek. Sodek had never driven before he came to Korea. He grew up in the city of Boston and said he never needed a car or had the desire to drive.

Of course, when he got to Korea the Army decided to teach him to drive. One of his first lessons was on the main road, outside the motor pool, which was mostly dirt and lined with rice patties. He was driving our typical five-ton bridge truck with a huge bridge bay section on the back. The bridge bay had these huge round prongs which stuck out on either side of the bridge section and were used to lock the bridge sections together. Sodek was driving along, and a bus was coming down this narrow dirt road right at him. He didn't get over far enough for the bus to clear those metal prongs and he opened up the side of that bus like a can opener! Luckily no one got hurt. He finally did get the hang of driving and turned out to be a good driver. But when E-109 was pulled out of the garage and the alert siren went off, Sodek was still in the early stages of learning how to drive.

So Sodek hopped into my truck and in a rush turned the ignition key to start it. The truck was still in gear, and it lunged forward and slammed into the huge garage door. Startled, he just sat there with a death grip on the steering wheel as the truck continued to lunge forward and into the garage door. After several lunges E-109 finally conked out and came to rest against the garage door, as a slow stream of antifreeze drained onto the motor pool ground. Needless to say, it was back into the garage for E-109. That wouldn't be the last time it was laid up.

KATUSA's were even worse drivers, probably because they never had the opportunity to get behind the wheel of a vehicle before having to do their mandatory service. In fact, they were not even allowed to sit behind the wheel of a vehicle in the motor pool when we did our morning vehicle warmups. This was due to an accident that took out the electrical grid for miles around Camp Pelham.

Two KATUSA's were sitting in one of the 5-ton bridge trucks and had it running to get it warmed up. The truck was backed up to the perimeter fence, and just beyond that was an electrical pole. Well, one of the KATUSA's, either by accident or on purpose, knocked the gear shift into reverse and the truck lurched backwards. It went right through the fence and took out the electrical pole, causing a major power outage.

We had been doing day and night bridge building training on the Imjin River. It had been cool and raining for the past several days and everything was muddy and the rice paddies were full of water. My truck held a bridge bay section and after the bridge bay was dropped into the water it was parked up along the dirt road lined by rice patties. I was working on the actual bridge construction so the evolution of driving and parking my truck was done by someone else. We worked all day and late into the night in the cold fall rain. Late into the night we started to wrap things up and head back to camp. I walked up the road through the rock road block, known as a tank trap, to my parked truck.

Tank traps are these huge areas cut out of the river bank and lined with rocks and explosives. If the North were ever to attack, these rock formations would be detonated and the road way would be covered in the falling rocks.

Tank trap in the background while assembling a bridge on the Imjin River

After reaching my truck late that evening with my buddy Tim Bailey we fell asleep awaiting the order to move out. We had quickly learned to fall asleep at any available opportunity. At some point after falling asleep someone came by and banged on the truck door and yelled to get ready, we were moving out. I opened my eyes and saw the small slits of light coming from the blackout lights of the company commander's jeep as it pulled up to the truck in front of me. At night we operated in blackout conditions and all you could see of the truck in front of you was the small sliver of light coming from its blackout lighting. I started the truck and when I saw the truck in front of me pull away onto the road, I put my truck in gear and started to move forward.

After only going several feet, the truck started to lean over toward the passenger side and slipped off the road and down into a rice patty. I was pretty confused about what had just happened along with my now wide-awake passenger, Tim. We quickly got our wits about us, climbed out the driver's side window, and onto the road. We could then see, through the pouring rain, that the truck was indeed lying on its side in a rice patty. One of the sergeants bringing up the rear told us to sit tight until morning and they would send someone back to get us. Tim and I got back in the truck, bundled up to each other, and slept until morning.

When morning came, we awoke to the sound of this huge M88 wrecker tank and a military tow truck coming down the muddy road. The M88 was basically a M1 tank with the turret and gun removed and a towing crane in their place. The M88 with its huge treads maneuvered around in the mud with ease and just got a hold of us and yanked us right out of there. Once out we were towed back to the motor pool.

The next day I was hounded with heckles from my peers and SFC Garrett, my platoon sergeant, wasn't happy at all! After lecturing me on how pissed the farmer was that his rice patty was munched up by an M88 tank, he said he was presenting me with his first ever "dog bone" award. The "dog bone" was actually the piece that held the bridge bay section together so I don't know why he was giving it out as an idiot award. I think he just liked the name of it!

After our work day we would scamper back to our barracks and start to drink.

We liked to "get our drunk on" before hitting the village where mama-san had the prices jacked up. We all liked to prove how much we could drink. It was sort of an indication of how manly you were. And it was common for us to get a case of beer and a bottle of liquor and polish them off before changing into our civilian clothes and heading into the village of Sunyu Ri. That was of course if you didn't have duty, were not restricted to camp, and you had a pass.

Me, Tim Bailey & Mike Johnson getting our drunk on

Passes were limited and they were only good until midnight. There was a curfew in effect up where we were and no one was allowed on the streets during curfew, military or civilian. The normal pass had a red stripe at one end and the overnight pass had a yellow stripe. You would go pick your pass up from the person on duty at the orderly room who had a list of people eligible to leave for that given day. When on pass we never strayed too far from the camp's main gate in the event of an alert.

As you went through the camp gate on pass you grabbed a few condoms - if you were smart - checked out the sign on which was posted the top five venereal disease (VD) clubs and made a mental note to stay clear of them!

Top VD clubs in Sunyu Ri and the nearby towns

Outside of most military instillations no matter where you are in the world, you'll find seedy bars and illegal activity. Korea was no different, only here prostitution was legal and accepted by the military and civilian community. And where there is prostitution there are sexually transmitted diseases.

Once out the camp gate and in the village, there was no shortage of bars and whore houses. The larger bars, like the Paradise and Blue Angel, were considered neutral ground and open to anyone. They were loud and usually had a lot of scuffles between engineers and gun bunnies, our name for the 2/17th field artillery guys. Most of the smaller bars were claimed exclusively by us, the engineers, or the gun bunnies.

You didn't want to make a turtle mistake and walk into the wrong bar or you could get your ass kicked. These bars were operated by an older lady known as a mama-san and she usually had three or four girls working as prostitutes. Their job was to make money for mama-san and themselves.

The whole prostitution thing, out in the rural behind–the-times Korean countryside, seemed weird to me but the Koreans seemed to behave like it was all very normal.

There were a couple of things you learned real fast when in the village. First, don't flash your money around. As soon as mama-san or the prostitutes saw you had some they were on you like a bad smell on a hog. Second, don't waste all your money buying the prostitutes drinks. They hounded you to do it because they made money for mama-san. You paid top dollar for a drink for them but it was really just water. It was better to just fork over the $10.00 or $20.00 USD and have them for a short time (quickie) or long time (overnight). Third, never drink from an open bottle. Mama-san would always try and put an open bottle of beer on the bar for you. You never accepted it unless you see her open it, or better yet open it yourself. Mama-san was a money maker and she was famous for pouring left over beer into a bottle and serving it up as a new fresh bottle. Her odds of pulling this trick off increased as you got drunker.

Mastering these things would hopefully move you out of dumb turtle status to just turtle.

Mama-san had no problem running a tab for her patrons and that usually kept a loyal group of guys in the bar all the time. At the end of each month, we would get paid in cash in the unit's orderly room.

Returning from a road march on payday

This of course was after we had completed a 10 to 15-mile road march in full field gear. And our platoon sergeant would check our ruck sack and canteens at the end to make sure we were carrying enough weight. We were always threatened that if we didn't make the march, or didn't carry the proper weight, we wouldn't get paid.

I made an extra $8 a month in overseas pay and had $250.00 sent directly to a savings account, leaving me with about $280.00 a month for spending. That was enough to get me into plenty of trouble!

Once I had received my pay from the pay officer, mama-san, Snap, and a whole list of characters would be waiting right there in the orderly room to collect their debts. Mama-san would usually ask in broken English, "You see Smith?" or Jones or whoever. Usually my answer was, "He go to land of big PX" and mama-san knew that he had transferred back to the states and stiffed her.

The prostitutes were in all the bars and we got to know the ones in the bars we frequented pretty well. We spent most of our evenings drinking, shooting pool, and hanging out with them. They were our age and at the time filling a legal profession. They had to be documented and checked for health issues on a regular basis. Sexually transmitted diseases were an accepted consequence of hanging around prostitutes. We were young, full of testosterone, and felt that nothing was going to hurt us, even VD.

If you did contract VD or any other sexually transmitted disease it was something you couldn't ignore. The pain and discomfort was unbearable and you would have to see the medic. First, in order to see if it's VD the medic would do a procedure called "rodding." We all knew about rodding but it was something you never wanted done. It starts with the medic taking a six inch long metal rod with a cotton swab at the end and inserting it about three inches up into your penis to get a culture. Now if you think that hurts, wait until you take a leak later. Oh my! After having that done you get marched down to the village and you are required to point out the prostitute you were last with so she can get tested and quarantined. Let's just say you're not welcome back in that club after you dime out one of mama-san's girls, took her out of commission for a few weeks, and probably put her club on the top five VD board at the camp gate.

Oh, and when you are out-processing to go back to the states, you are "rodded" again just to make sure you're not taking anything home.

VD was a huge heath problem in Korea at the time and there were articles in the American servicemen's paper called the "Stars and Stripes" that said there were untreatable strains of the disease floating around the country. In an effort to come up with a drug to prevent VD the local clinic was soliciting volunteers to participate in an experiment to test a new VD vaccine. Some would get the real vaccine and others would just get a water solution. But you would not know what you received until after the study. The whole idea of the Army testing a drug on me was not appealing, and I opted out. A few of my friends did participate, got the real vaccine, and their feelings were mixed on the results. As far as I know the vaccine was not a success.

Speaking of the clinic, I was over there one day waiting out in the tiny waiting room with another guy who was from the 2/17th. He seemed nervous and was sweating and visibly shaking. I asked him casually how he was doing and he said, "Not to good, I accidently injected myself with atropine." During this time, and through most of my military career, we were in the cold war with Russia. And one of the biggest threats we prepared for was their use of nerve agents against us. The antidote we carried for nerve agent was atropine and 2-PAM-CL auto injector shots. If you felt you were exposed to nerve agent you would administer these shots, or injectors, into your thigh. You would use your thigh because the needles on these things were pretty long. Once you were done you would bend the needle and hook it through the collar of your shirt so if you passed out someone coming along would know how many shots you had given yourself, and wouldn't over dose you with more.

You didn't want to take one of these injections unless you had to. The side effects are just about as bad as nerve agent without killing you!

You get nervous, anxious, confused, hallucinate, shake, and vomit; get the picture?

Anyway, this poor guy had somehow shot himself with one of these injections by mistake, horsing around or trying to get a buzz. Medical just had him sitting there while they watched him until the effects wore off.

After a while in country some guys would find themselves a Korean girlfriend and rent a room, known as a "hooch," in the village. These girls were usually prostitutes still working for mama-san or ex-prostitutes who had gone out on their own. Some of these relationships blossomed and it wasn't unusual for guys to get married to a local Korean.

The normalcy of having a relationship drove me to seek out a girlfriend. I ended up meeting a girl who was waiting for her previous GI boyfriend, who had transferred to the states. She was waiting for him to send for her and she would be moving to the states when he did. That worked for me.

She already had a hooch which was a typical one room simple structure that led into a courtyard surrounded by other hooches. The hooches all shared a common hole in the ground toilet that you squatted over and it stunk. There was no running water and all your washing and bathing was done down at the bath house. I never ventured down to the bath house, choosing to use the facilities on camp instead.

The hooch was heated by placing coal blocks in a heater outside known as an ondol, which heated the hooch floor. The older man who rented my girlfriend the hooch took care of that. I asked him what he wanted in payment for tending the ondol and he requested razors for shaving. So, whenever I would see him I would give him a bunch of razors, and boy would that put a huge smile on his face!

My girlfriend didn't last long. Overnight passes were limited and I only got one every couple of days. The first time I had one I over slept my alarm and was late getting back on camp for morning muster. SFC Garrett placed me on a week's restriction and extra duty. In those days there was no paperwork for minor infractions, your platoon sergeant took care of you. Extra duty consisted of breaking down semi-truck tires in the motor pool after work. There was never a shortage of tires because SFC Garrett would have them stacked up for just this type of situation. It took hours to break them down using a sledge hammer and wedges.

SFC Russell A. Garrett

The next time I had an overnight pass the same thing happened. I was late for muster because my alarm didn't go off. SFC Garrett gave me more of the same punishment. While I was breaking down tires SFC Garrett came down to see me. He was a rough and tough older sergeant who always seemed to be chewing tobacco, the type in the pouch that produces that big spit. I knew he was married to a Korean and this was his second or third tour over here. He was a good, fair man and gave me some fatherly advice; get rid of the girl! He told me, "You know that girl is playing you. She is messing with your alarm so you over-sleep and get restriction. She knows how the system works. With you on restriction she has the place to herself and does what she wants while you pay the rent."

Good advice from the old sergeant to a naive 19-year-old, although I was enjoying the female companionship. Even though it was an awkward situation, it seemed better than paying the prostitutes. But when I was off of restriction I went and got what little I had in the hooch and was done with that.

In early March we headed south for a huge war game known as Team Spirit. This was a joint exercise with the U.S. and ROK military and would be the largest joint exercise in the free world. It initially kicked off around the 1st of February and would last until the end of March, involving more than 191,000 Korean and American military personnel. It would involve every aspect of warfare to include the Navy's 7th fleet, two carrier battle groups, marine amphibious tasks force, Air Force Strategic Air Lift and Tactical commands, and several Army divisions to include the arrival of the 7th Infantry Division from California and the 25th Infantry Division from Hawaii.

Of course, the exercise gets North Korea all worked up. They go on a proactive heightened state of tension and look at Team Spirit as an act of aggression. The North announced it would be in a semi-war state of alert for this year's event, the first time they had reacted so strongly in the eight-year history of the exercise. Remember earlier when I mentioned the North Korean Mig-19 pilot who defected in February? He claimed he defected because the North Korean Communists were pushing war preparations in a frenzied manner, claiming war was the only way to unify the divided Korea. He said things escalated as the South started preparing for Team Spirit this year. The atmosphere was so tense at Camp Pelham I was actually happy to be deploying south for the war games and away from the DMZ.

I would be driving my truck to the operating area and my passenger on the trip would be Terry Bear who had been in boot camp with me. Terry was a happy-go-lucky guy who always had a cigarette in his mouth and would mumble when he talked. He was a good guy to make the trip with. We would be driving over eight hours south of Seoul to where the war games would take place. This would not be an easy drive with our huge equipment on the twisting old Korean roads which were used by outdated Korean farm equipment. On the way down one of our vehicles got into an accident with a Korean pedestrian and there was a fatality. The incident had us stopped for hours waiting to clear the matter up. It was sad that someone had to die. You knew it was bound to happen with our huge vehicles on the tight confined country roads. Thankfully the rest of the trip was smooth and punctuated only by several stops to camp overnight.

Me & Terry Bear Heading to Team Spirit

Once down in the Team Spirit operating area we became part of the Blue Forces that would be fighting the Orange Forces. We were identified by blue bands around our helmets and blue triangles on our vehicles. We initially dug into a huge sandy field and covered all of our gear with camouflaged netting in preparation for the upcoming battle.

This was March and wouldn't you know it, it snowed! It was actually forecasted to be a colder than usual winter and if it wasn't snowing it was freezing rain. We were warned to keep on the lookout for signs of frostbite and trench foot. This was camping at its best with below freezing temperatures, outdoor shitters, tents, and canned food officially known as Meal Combat Individual (MCI), known to us as "C" rations.

There were no showers and you washed and shaved out of your helmet. As usual with being in the field, we would get our only hot water, a helmet full, at breakfast. Our sleeping quarters were the usual huge green GP medium tent.

Our tent & latrine under camo netting

We would all pile into that tent which was heated by a gas stove and hunker down for the night. Our biggest enemy was the cold. Our second concern was theft. Out here in the scrub you always kept your gear away from the tent edge because there may be sticky fingered locals about known as "slicky boys" looking for their opportunity to swipe our gear.

We didn't worry about the actual enemy, the Orange Force, because we had a couple of M60 tanks dug in around us keeping us safe.

M60 Tank looking over us

Our portable toilets were made of a wooden board with holes cut out of it which sat on 55-gallon drums cut in half. The only thing that kept you from being seen sitting there doing your business was the camouflage netting draped over the thing. Once a day, in the morning, someone would get detailed to burn the waste in the 55-gallon drums, never a good detail.

Even at this remote location the locals eventually found out where we were. They would drift in to see what we were up to and "ahjussi," a Korean older male, or mama-san, would be selling cooked noodles or soda out of his or her makeshift kitchen that they carried on their back. Even in the rural country side there are entrepreneurs!

The battle started out in the early morning hours around the 23rd of March with Orange Forces crossing the Hwachon River into Blue Force territory on a major offensive. The battle picked up from there as we waited in reserve until a river crossing was needed by our Blue Forces. While waiting we found a nice sized lake and commenced ferry, or what we called rafting, exercises to hone our skills. The lake was really different from our usual Imjin River site. The lake was bordered by stones, which were harder to drive on, and easier to get stuck in while backing up to the water to drop a bridge section or boat. The training kept us busy and we were eager to get the call to head to the front.

In this next set of pictures, you can see a bridge bay section, this one is a ramp section, being launched into the water and then picked up by a 27' boat. From here it is taken to be assembled with other sections to form a raft or bridge. The boats are dropped in first, then bay & ramp sections.

Dropping a bridge bay section during Team Spirit 1983

After about a week we packed up and moved out to the battle front. The Blue Forces were on the move and counterattacking the Orange Forces. On the second day of the counterattack, we were called up to build a bridge to facilitate a river crossing of the Han River, at Ipo near Ipo-ri, about 75 miles south of Seoul. The only noticeable landmark on the way was the OB beer factory near Icheon.

When we made it to the Han River, about the 1st of April. The 44th Engineers (Combat Heavy) had built a causeway jetting out into the river to provide a bridge head and a huge smoke screen was released to cover us from enemy forces. This was created by a unit with huge smoke generating machines that belched smoke at just the right concentration to allow us to do our work and yet remain invisible to enemy forces. Under the smoke screen we deployed our bridge sections and boats into the water and started ferrying elements of the spearhead attack force across the river. The ferry was made by connecting several sections of the floating ribbon bridge and securing one of our 27' boats to each side. The ferry, or raft, commander would then stand at the front of the ferry and direct the boats with hand signals to guide the ferry across the river. Lead elements of ROK Army cleared mine fields left by the Orange Forces on the other side under a cover of tank and mortar fire.

Raft configuration

Once we had the advance force across and we had secured the beachhead and buffer zone, we commenced to build a full crossing bridge from one side of the Han River to the other. This was in place long enough to cross major tank and armored personnel carrier (APC) elements of the Blue Force onward to engage the Orange Forces. It was pretty cool once the Blue Forces were rolling across because a trio of B-52 bombers from the Strategic Air Command came roaring overhead in a simulated bombing support run of our advancing forces. What a sight that was.

Our bridge during Team Sprit 83

When the bridge was no longer needed, we packed it up and commenced following the Blue Force column in case they needed to cross another river. We were getting into April and the weather was turning to cold rain and mud. The muddy roads were getting especially bad after following the tracked tanks and APCs that made a mess of everything. Soon after packing the bridge up our long convoy had come to a halt along a dirt road near a dried river bed. As we sat there waiting to move out, I could hear this low thumping noise in the distance. The noise kept getting louder and louder until I could see about 20 or 30 Huey helicopters coming straight for us. I was sitting there in my truck with a front row seat to this spectacle thinking, wow this is pretty cool, never thinking it could be the Orange Forces attacking us!

View from my truck window of the Blue Force air assault landing in the wrong LZ. You can see our command "gama-goat" vehicle giving directions.

The helicopters came in fast and low and landed in the river bed next to our line of parked trucks. They came in to land in waves of about five and as soon as the helicopters hit the ground, soldiers loaded down with gear piled out of them and started setting up defensive positions along the dried river bed. The helicopters relieved of their cargo took off to allow the next wave to come swooping in. After the first wave hit the ground I noticed our commanding officer's vehicle, an ugly half jeep and half trailer known as a "gama-goat," pulling down to what appeared to be the helicopter air assault force leader. There was some discussion and a few minutes later the empty helicopters came swooping back down to pick up their cargo of troops. It later filtered down the grapevine that the air assault guys in the helicopters were Blue Forces. They had landed in the wrong landing zone (LZ) and needed to be a few miles ahead attacking Orange Forces.

At one point toward the end of the war games my platoon had the unusual opportunity of harassing the Orange Force. I guess, since we had finished our bridge duties and had nothing really left to do, we seemed like a logical choice. We were sent up toward the Orange Forces lines for a few days and pitched our GP medium tent. It was cold out and the ground was frozen solid. Unlike our strict code of military dress, such as helmet on at all times and uniforms worn per regulations, we would be dressed on this mission anyway we wanted. Since all we really had was our uniforms, the only thing we really changed was swapping out our helmet for a black ski mask cap. It was an unbelievable relief not to have that damn helmet on for a few days.

While conducting this operation we would be using a shooting system that sent a laser beam out the barrel of your weapon every time you fired a blank shell. The system was fairly new and known as Multiple Integrated Laser Engagement System (MILES). Everyone had vests on that had sensors on the front and back. If you were hit by the laser shot from someone's weapon your vest emitted a constant beeping that sounds like a smoke alarm, indicating you had been shot. If it only beeped once it meant someone had taken a shot at you and it was a near miss. The only way to stop the constant beeping was to take the key out of the laser box on your weapon and put it in the box on your vest. Doing this disabled you from firing your weapon and took you out of the game.

To me, the most memorable event of the operation had us going out at night to ambush an Orange Force patrol. We made our way up a hillside overlooking a road that was about 25 yards away. We had a perfect view of the road below us and we set up about five yards apart from each other. We would wait until we had the bulk of the patrol right between us before opening up on them. Shortly after we set up the orange patrol began to slowly and quietly come walking down the road in a typical staggered formation of two lines. We waited as planned until the bulk of them were right in front of us and we opened up with M16 fire. All you could hear was the pop-pop of the M16s and the beeping of laser gear going off! When we stopped firing and turned to get out of there the sweet sound of about 20 lasers beeping let us know we had accomplished our mission. We headed back to our GP medium, slipping and sliding all the way down the frozen road, laughing and telling tales reliving the skirmish!

As the war games wound down, we started to free-lance and the company drifted off into the countryside to do separate bridge training. We had come to this river lined with rocks and it was decided that this would be a good area to try and do some ferry training. A scout jeep was sent out to see how fast the river flowed, how deep it was, and the best possible place to cross. After sitting around waiting for a bit I walked down to the river's edge and asked a guy standing there where the scout jeep was. He pointed out to a sand bar in the middle of the river with a yellow flashing light just below the water and said nonchalantly, "It's under that flashing yellow light." After that it was decided we would camp in the nearby field until they figured how to get the jeep out of the river.

As we set up camp a young Korean had come down from the nearby village and was talking to one of our KATUSAs. The young Korean was studying English and wanted to hone his skills. He told the KATUSA he wanted to invite one of the Americans up to his family's restaurant for dinner and to speak English. The KATUSA asked me if I was interested, and I said yes. After clearing this with my sergeant I walked with the young Korean into the village dressed in my field gear along with my M16. We arrived at his family's restaurant and had a seat at a large table with a traditional gas grill built into the middle of the table. The young Korean and I talked in English as his family joined us around the table for the family style Korean meal. As the propane gas grill heated up, food was placed on it and once done was picked off for consumption.

After about 30 minutes of sitting there I started to feel really dizzy and my vision started to blur. All I could think of was that these Koreans were trying to drug me and get my weapon or kidnap me! I started to get up and stumbled. I made it to the door as the young Korean and his family were yelling and getting excited. This just fueled my concern for my safety, and I ran out the door and stumbled back to the camp in a daze. Once there I explained what had happened to the KATUSA.

Soon after my arrival the younger Korean followed, and he and the KATUSA discussed what had happened. It turns out that there was a gas leak on the stove, and I was starting to be overcome by the gas fumes! I was embarrassed and apologized to the young Korean. The KATUSA affirmed my apology to the young Korean just in case something I said was lost in translation.

Team Spirit ended, and we returned to Camp Pelham. I think this picture of John Fisher in the motor pool after returning sums up how we felt.

John Fisher after returning from Team Spirit

A few weeks later we were standing in formation and the company commander presented us with "Team Spirit 83" key rings. That would be our reward for over a month in the mud, rain, snow, and freezing temperatures conducting a war game.

In May I was promoted to Private First Class (PFC) and got a $25 pay raise. There was the official "pining on" in front of the company by the commanding officer, and then there was the unofficial "tacking on" by my platoon. This latter ceremony was done down in the motor pool in private. The "tacking on" was accomplished with the platoon standing in two rows facing each other. I then had to slowly walk down the middle of the two rows and receive a punch on the arm from everyone in the platoon. Some guys just give me a light punch, but others hit me with everything they had.

On one occasion a KATUSA was promoted to corporal. He was a small shy guy and was reluctant to walk down the "tacking on" gauntlet, so he ran through it. About halfway down the row of guys, someone caught him right in the jaw with a punch. It ended up dislocating his jaw!

Not long after we returned from Team Spirit, I ran in my one and only half marathon, which was 13 miles. It was a huge event held on Friday the 27th of May and involved the whole battalion. The 2nd engineer battalion was headquartered out of Camp Red Cloud, which was located in Uijongbu, about 20 miles southeast of Camp Pelham. The battalion had just started this tradition the previous year and it involved the five subordinate companies consisting of the Stallions, Beasts, River Rats, Village Rats, and the Renegades. These companies were spread out over the DMZ area and this was the only time we ever got together as a battalion.

The half marathon, which finished 2-1/2 hours later at nearby Camp Casey in Tongduchon, was slow and painful. Because we ran in such a large group, 552 soldiers, the pace was more like a fast walk than a run and I just couldn't get into a running groove. And of course, it rained. So, we were all soaking wet at the end. We looked like "Drowned Rats" instead of "River Rats." The only upside was at the end of the run, there was a picnic complete with beer. Oh, and there was a commemorative t-shirt that read "CRC in 83." I had only been in Korea for six months and had been awarded a key ring and a t-shirt!

Soon after the half marathon our company's leading sergeant, known as the first sergeant (1SG), began soliciting for volunteers to attend the week long ROK 9th Infantry Division Ranger Training Program. The 1SG was a "no bullshit" type of guy by the name of Sauceda, but we always addressed him as "first sergeant."

This was an elite branch of the ROK Army and similar to the U.S. Army's Rangers. I had seen the ROK Army in action plenty of times and knew I didn't want anything to do with them! And yet, for some reason, I ended up volunteering along with 10 other guys like PVT Mike Johnson, PVT Neil Swanson, PFC Mike Geurtin, SP4 Jeff Hanks, and several KATUSAs like CPL Kwon. Maybe I thought it was like back in boot camp and we would spend the week sitting around a lake water skiing! Well, 1SG Sauceda was one tough guy and he knew we were not going to be water skiing. And he was not about to have himself or his men shown up by these ROK Army Rangers. For nearly a month before we left for the training 1SG had us running and doing exercises twice a day. We would do our normal PT in the morning and do our ROK Ranger training in the evening. This evening training was nothing like our normal PT. 1SG had us doing long runs in the back alleys and along the narrow rice patty paths, so we had to pay attention while we ran. It was good preparation, but it soon became apparent that it wasn't nearly enough.

We left Camp Pelham in late June. We headed out into the deep countryside about 20 miles away in our company's deuce-and-a-half to the Kamaksan Training Compound near Uijeongbu. After winding through the hills and into the middle of nowhere we arrived at the ranger school. There was really nothing there except for a simple concrete building which was their main office and mess hall. We would be sleeping in our own two-man tents, eating our own "C" rations, but sharing their bath house. The bath house wasn't a house at all; it was open and consisted of a couple of pipes sticking out of a natural rock wall about head high with spring water flowing out of them. The water was as ice cold as ice cold could be! The toilets were the traditional hole in the ground; you just straddled the hole and did your business.

We got our gear together and started up the hillside to our camp site on the side of a good-sized Korean hill. Once at the camp site we merged with several other American units participating in the same weeklong training. The first couple of days were filled with non-stop obstacle courses and running.

Initially we started with a course that I assume was intended to see what kind of shape we were in. It was basically a long obstacle course running through the woods along the mountain side. You would be running through the woods on a narrow path and suddenly come to a huge obstacle. You had to clear it or keep trying before you could proceed. It was only the first day and 1SG was getting upset at the ROK cadre because he thought that they were going a little too far with their discipline. It seemed to come to a head when we were trying to clear this one obstacle, which entailed swinging from rope to rope over a huge mud puddle. Some guys were beat-tired and could not get out of the mud once they fell in. The red shirted ROK cadre had no tolerance for failure. When someone would fall into the mud, they wouldn't let them out and kept pushing them back into the mud as a sort of punishment. 1SG had one of the KATUSAs reluctantly translate that enough was enough. After some testosterone positioning on both sides the cadre seemed to give us a little slack, for that obstacle only.

Catching our breath between obstacles

Needless to say, the first day was long and we were all beat. There was no getting out of the spring water shower at the end of the day. We would be covered head to toe with dirt and mud and just wanted it off. The initial shock of the ice-cold water never did subside, so you got wet, soaped up, and rinsed in a hurry. I would then eat my "C" ration and fall fast asleep. Luckily there was no guard duty at night.

After the initial obstacle course, we advanced to riskier and riskier obstacles like rappelling down a hill's rock face, crossing valley gorges on single strands of rope, pulling ourselves up 50-foot towers and then rappelling down them, just to name a few. They were all scary. I mean the ROK Army trained in a realistic manner. Injuries and death during training were all part of the deal when you engaged with them. They experienced danger every day from incursions from the north. So, they trained for real life threats.

Rope Crossing Obstacle

Rappelling down the hillside

Rappelling down the hillside's steep rock face seemed the most dangerous to me. Partially because we had to pull ourselves up 100 or so feet of almost vertical hillside by rope and partially because we were jumping off of it holding onto nothing but a rope. But it seemed the most dangerous because once we were at the top of this steep hillside, the cadre had us shimmy out on this long narrow rock ledge where we waited in line to be hooked up to rappel back down. There were no safety lines or anything else to prevent us from falling as we waited in line. We were all clutching the side of the hillside in fear of falling to our deaths! Meanwhile the cadre casually maneuvered around our frozen bodies, hooking us up and sending us on our way.

ROK cadre in red shirts and Echo company trainee's

Toward the end the young ROK Ranger cadre softened up and we took the time to take group photos and share some laughs with the KATUSAs translating for us.

Our last obstacle was an extremely long zip line that started way up on the mountain, ran just above the tree top canopy, crossed a large pond, and ended near the compound. In full gear I grabbed onto the handle and was released. It didn't take long to build up speed. As I zipped on down the mountain, I couldn't believe how at peace I was, finally being alone for a few minutes as I flew down that mountain. By this time, I had enough confidence to not be afraid of anything.

ROK Ranger Patch

On our last day, we got as presentable as possible and formed up in the gravel courtyard in front of the main ROK building on the compound. As we stood there at attention, I was presented my Ranger patch and certificate from Major General Ko Myung Seung, the Commanding General of the ROK's 9th Infantry Division.

I was extremely proud to receive my ROK Ranger patch and certificate after completing the course on the 2nd of July. Unfortunately, the patch was not authorized to wear on my uniform. I have kept it in a safe place over the years and have it proudly displayed in my Army shadow box of medals and ribbons.

Loading up for ADA duty. Left to right; PFC Semler, PFC Bailey, SGT Groceman, PFC Cotton, PFC VanAuken, CPL Joo, PVT Houston, PVT Johnson, PFC Fisher, PVT Paul

In late July 1st platoon was detailed to support the 2nd Battalion, 61st Air Defense Artillery (ADA) in an exercise. They were stationed right on the other side of Sunyu Ri at RC4. The live fire exercise was going to be held at Chulmae Range near Daechon Beach, which was southwest of Seoul and on the Yellow Sea. This was over 150 miles away and most of the trip would be by train. The members of the platoon going were SGT Groceman, PFC Lewis Blankenship, PFC Tim Bailey, PFC John Fisher, PFC Milo VanAuken, PVT Mike Johnson, PVT Jerry Paul, PVT John Cotton, PVT Brian Houston, me, and our KATUSA CPL Joo.

We drove the truck to the nearby Munsan train station and had it loaded onto a flatbed car along with the air defense vehicles. Of course, we had to be different and added a red "River Rats" flag to the front of our truck.

Train Ride to Daechon Beach

After making sure the truck was safe and sound, we took our passage in one of the regular train cars.

The train ride was slow and rambled through the small villages and countryside. At one point we stopped at a large station and got off to use the bathroom. It was at a major train hub and there seemed to be a lot going on. The bathroom was a huge room with stalls but no toilets or urinals, just holes in the ground for you to relieve yourself in. Oh, the smell!

At Daechon Beach we off loaded our truck and drove over to the live fire location at Chulmae Range, which was a U.S. military complex.

Once there we set up our gear in another huge GP medium tent, only this one had metal bunk beds in it. That was all that was in the tent, rows of metal bunk beds. It was warm so we had the flaps of the tent on either end tied open to catch the breeze. We thought since we were on a military installation, we could leave our gear lying about. Wrong! Even though it was a military complex slicky, boy got my cassette player radio one evening while I was asleep. Slicky boy was everywhere!

Over the next few days we could tell that other things were turning up missing and that someone was getting into the tent during the night and stealing. Private Houston also had a cassette player, and he was worried it was going to be next. He devised a plan of placing empty beer cans all over the inside of the tent, as a sort of booby trap. When slicky boy came into the tent in the middle of the night, he would be bound to kick one of them over and wake us up.

Sure enough that night I awoke to a pitch black tent and all hell breaking loose! Houston and several others had been awakened by the kicking over of a beer can and they had wrestled down a Korean there in our tent. As soon as the Korean saw Corporal Joo he kept saying he was a friend of his and he was just coming to visit. Corporal Joo however said he didn't know him. We turned the intruder over to the camp military police who were Korean.

The next day those of us who were missing items were called to the camp's police station to identify and pick up our missing belongings. My cassette player was there, and I regained possession of it. I also felt bad for the slicky boy who I knew was probably getting the shit kicked out of him somewhere. The Koreans are a very honorable society, and they were shamed by the event. And I was sure slicky boy was paying the price for embarrassing his fellow countrymen.

In the evenings we would put on our civilian clothes and walk into the sleepy seaside village for a few beers and walk the beach watching the villagers fishing or hunting for clams. Daechon Beach was a seaside resort and very relaxing compared to the constant tension up on the DMZ.

Chaparral missile APC

We would spend a week here as the ADA guys fired Chaparral, 20mm Vulcan cannons, and Redeye missiles at unmanned drones. Now the drone part is where we came in. We had a rubber commando style boat with an outboard motor on it and we sat on the beach and waited until the ADA folks shot down one of the drones. When they shot one down, we would hop into our boat and speed out to the drone and retrieve it. If they missed the drone, it would be directed close to us by remote control and the engine killed. A parachute would then deploy, and it would float to the water for us to retrieve. Not bad duty!

Our small boat and a drone that didn't get past the live fire

It was pretty interesting watching the ADA process unfold. The small APCs that the chaparrals and 20mm Vulcan cannons were mounted on would line up at the beaches edge on an elevated platform overlooking the Yellow Sea. The 10-foot-long red drone was pulled by a remote control plane that was launched from a huge catapult. When they were ready to fire, the plane was catapulted off into the air and the drone was pulled off the ground behind it. The ADA guys would then track it and fire their weapons.

20mm Vulcan APC

This wasn't a cheap event. The Chaparral missiles went for around $70,000 per round and the Redeye missiles were around $25,000 each. At those prices you want to get the best training with a limited amount of misses.

After they fired, we would do our thing and go get the drone or wait for it to float down to the water. The ADA guys fired day and night, which kept us pretty busy. During this exercise they made history and conducted the first night firing of a Redeye shoulder fired missile in Korea. The following picture is of that event and taken from the "Indianhead" newspaper, the official paper of the 2nd Infantry Division.

(L-R) Sgt. Nelson Nathaniel and PFC. Michael Odum, of Battery C and Sgt. Charles W. Seldon, an instructor at the range, practice tracking a target before Korea's first night firing of a Redeye missile.

When we finished up at the end of our week-long tour we loaded our truck back on the train and headed back to Camp Pelham.

In the fall the company geared up for another exercise up around Camp Pelham. We wouldn't be taking our bridge equipment this time because this training would be to hone our infantry skills. Conducting infantry training meant long days and short nights with little sleep, basically pretending we were at war. So, we loaded up and headed out into the countryside.

We set our company perimeter up and dug foxholes per our normal procedure. When we set up a perimeter in the field there are two guys to a fox hole. At night one of the two was awake and on watch while the other slept. We usually rotated one hour on and one hour off.

I was getting ready to fall asleep one night and the sergeant came around and said I had been detailed to go out to a forward listening post about 150 yards out from our perimeter. Great, I'm tired as hell and now I have to go pull some duty. John Fisher and John Cotton would be sent out with me. The three of us walked over to the command tent, picked up our radio, and were briefed by the lieutenant. He told us that if we see or hear anything we needed to radio it back to him. With a smart "Yes sir!" we headed out to the listening post area.

Now this was a little unsettling to me because up here on the DMZ the ROK guys did not mess around. They carried live ammunition and shot first and asked questions later. I mean these ROK boys were on war time standby 24/7, and as I said before, they routinely ran into North Korean infiltrators.

During this exercise we didn't carry live ammunition. I could never understand that. It was almost like the command was afraid we would hurt ourselves with it! That's probably why the last place I wanted to be was here, hiding out on this hillside and those ROK boys thinking I was an infiltrator. Cotton and Fisher felt the same way.

Nevertheless, the three of us made our way out to the listening post area and sort of just hid on the ground in the high weeds together side by side. We were in a ravine with steep hills all around us. We had a radio and were to contact the company command tent if we heard anything.

It was getting dark, and we didn't really have good bearings but knew that our company was behind us.

As we sat there, Cotton decided it would be a good idea if we had some sort of booby trap set up so that if anyone came up to us, we would know. Well crazy ass Cotton, who was already famous for getting stoned drunk one night and tearing up the gun bunnies' bath house, slipped out into the dusk of evening to find booby trap material. About 15 minutes later he came back with all this wire, which we thought was strange. Where would he find wire way out here? He said let's tie it to the small trees around us and it would act as trip wire. It sounded pretty good to Fisher and me, so we did it. Heck, we caught the slicky boy there at ADA live fire with beer cans!

We settled in laying there side by side in the tall grass and decided we would rotate watch and set up a schedule with Cotton leading off. At some point in the middle of the night I woke up to the sound of Korean voices and assumed they were ROK soldiers sneaking around. I looked over at Cotton and Fisher and they were both asleep! I looked to my side without moving a muscle and could see by the light of the moon; Koreans in uniform walking very slowly by us. I didn't make a sound and I'm not sure if they noticed us or not. I couldn't be sure if they were South Koreans or North Koreans! I lay there still as could be until I guess I fell back asleep because I was awakened by the light of morning.

I relayed the story to Fisher and Cotton about the Koreans walking by us in the night and asked them if they had seen or heard them. Neither one did and said they must have fallen asleep. They got very close to us, but not close enough to set off the trip wires.

We worked our way back into the perimeter and checked in with the command tent. After we turned in the radio, we briefed the lieutenant in charge that we had not seen or heard anything. He looked at us skeptically and asked, "Are you sure you didn't see anything?" We said "No, we hadn't seen a thing." If we told him we had, we would have had our asses handed to us for not calling it in. The lieutenant said, "That's strange, because the ROKs are all upset because someone cut a bunch of their communication wire last night and they had to run new stuff all around us in the middle of the night!" After that statement he told us to get back to our platoon.

The next several days were filled with sleepless nights standing guard in our foxhole and maneuvers during the day. One of these maneuvers was a 20-mile road march with all our gear. Road marches were not unusual as we did them about once a month. But being out here in the field with no shower and already being tired made this one worse to begin with.

Road marches were a good paced walk when on a road or path, and slow and sluggish when going through woods or fields. We walked on the sides of the dirt road, guys on either side, staggered, with about thirty feet between each other. As we walked along in silence, we passed around the heavy machine gun, known as the M60, taking turns lugging it. Every now and then the guys in front of me would stop or kneel down and I would do the same. After a few minutes we would be up and walking again. That's how it was done for 20 miles.

The following picture is Corporal Shin, a KATUSA, lugging the M60 on a road march.

Corporal Shin – a KATUSA – with our M60

But I must admit, I saw a lot of interesting things walking through Korea.

As we marched through the countryside we crossed this field of solid white feathers. I couldn't figure out what they were doing there until we got to the other side and there was this huge chicken processing building. We had just walked through the remains of millions of chickens!

As we made our way through a small village we encountered a couple of men under a bridge burning the skin off of a couple of dead dogs. They had these two dead dogs strung up under this bridge by their hind legs and were burning the skin off in preparation for a Korean meal of dog meat, which is a delicacy I'm told. At about this point into the march I could have helped them eat that dog I was so hungry.

The march drug on for most of the day with no stops to eat. Toward the end of the march we cut through a potato patch. I was so hungry I scooped one up and started to eat it. It seemed like a good idea. What an awful taste! I will never forget the taste of that dirty raw potato. Or the feel of the dirt caked in my dry mouth. Soon after getting back from the road march someone took a picture of me and my foxhole buddy next to our camouflaged foxhole. You can see I was not happy as I'm flipping the photographer the bird!

Taking a break in front of our foxhole after returning from the road march. I'm on the left.

That foxhole and I became pretty close during this exercise. I spent half the night there standing guard on my field of fire. Every morning at sunrise I was in it with my foxhole buddy to thwart a possible enemy attack – because someone figured out that was the best time for one. And where else can you get some peace and quiet, but in your foxhole.

The maneuvers were coming to an end just in the nick of time. We were all very tired and getting on each other's nerves. It didn't end fast enough for Cotton. He had some sort of medical breakdown and had to be medevac'd out by a helicopter, which they called a "dust off." After popping a smoke grenade this Huey medevac helicopter came swooping down in a cloud of dust and took Cotton away.

Apparently, Cotton had been taking these pep pills known as "skoshy yellows" and had over done it. Tim Bailey said he was riding in the back of a duce-and-a-half with Cotton when he started to go into almost a seizure type convulsion and was choking on his tongue. Tim said he tried to cut off the bandana around Cotton's neck with his knife, thinking that was causing him to choke, but Tim made things worse by cutting Cotton's lip! Eventually they got him out of the truck and the medevac helicopter was called in.

Cotton was always into something. He was such a good-natured guy and he loved to tie one on. Like I mentioned earlier, Cotton had gotten all messed up and ended up tearing up the gun bunnies' latrine. The story was that he went into the village, tied one on, tore up a club, and must have had a bad run-in with the gun bunnies. When he got back to camp, he decided to get back at them and took it out on their latrine. It caused a big ruckus and both commands got involved, which is never a good thing.

This snowballed the next day when Tim Bailey and I went into the village to do some drinking. We didn't know it, but we were heading into the club that Cotton had just trashed the night before. As we headed toward the club, which was down an alley off the main street, I told Bailey I was going down to RC4 to pick something up from the PX and would be back in about 30 minutes.

By the time I got back to the club Bailey was about ready to pass out. I shook him and asked him what was going on, but he could hardly speak. Just 30 minutes before he was sober as could be. I knew something was wrong, so I picked him up, put his arm over my shoulder, and out the club we went, with me pretty much dragging him down the alley. We ran into a few other guys I knew, and we got him back to camp and into his bunk. Later, we figured that mama-san must have recognized Bailey as being a friend of Cotton's and slipped something in his drink. Perhaps they were waiting for him to pass out so they could kick the shit out of him.

It wouldn't have been the first time something like that had happened. Obviously, they thought he was alone and didn't expect me to come walking in. When Tim finally came to, all he could remember was walking into the club. That was it.

During the winter of 1983 there was a plan to put a guide wire across the Imjin River to help us build an ice dam which would make a natural ice bridge. But this entailed drilling a hole in the cliffs above the river to run the guide wire through. At some point during the drilling process the expensive drill bit got stuck in the ground. Because the Koreans were known to be very slick with recovering anything for money, it was decided that a guard would need to be placed on this bit until the ground thawed and they could get the bit out. You had to respect the slicky boy!

I thought it was a bit much to think they could get the bit out of the frozen ground, but they did almost get a whole jeep out of one of the camps. Turns out they pulled a jeep into the back of a garbage truck, filled the truck with garbage covering the jeep, and tried to drive out of the camp gate. Luckily the guard was on the ball and probed the pile of garbage with a stick and kept hitting something hard. When they removed the garbage, they found the jeep.

Reminds me of a story my dad would tell when we were living on the Philippine Islands in the mid-1960s. We were living on Clark Air Force Base, and he said the locals stole the base fire truck. They just turned on the sirens like they were heading to a fire and drove right out the main gate with the gate guards holding the gates open for them. Guess slicky boy really was everywhere.

Getting back to the drill bit stuck in the ground, I was detailed to go out and guard this thing with another soldier. The truck dropped us off in the middle of nowhere on the cliffs of the Imjin River to an awaiting GP medium tent. There was a long metal shaft sticking out through the top of the tent about 20 feet in the air. Yep, this thing was that long.

The guards we relieved passed the usual guard duty log to us and were happy to get into the truck to take them back to nice and warm camp. I stepped into the tent and there was the rest of the metal shaft going straight into the frozen ground. There were also two cots pushed up against a gas burning stove – that's all that was there. My partner and I spent the day keeping the gas stove running and staying warm in the tent. The stove ran off of gasoline cans positioned outside the tent and we would just switch the suction line whenever a can ran dry. At some point in the middle of the night I woke up freezing cold and went out to switch the gas suction line only to find out our gas cans had been stolen! Slicky boy had struck again! Needless to say, it was not a good night for me and my frozen partner. Waiting until the morning when the relief truck arrived seemed like an eternity!

Cable holding back the ice to form an Ice Bridge

During my tour I did make it into Seoul on my only three-day pass. We caught the train in Munsan and took it into the heart of Seoul. The train was packed with locals, and we stood out like sore thumbs. We did some sightseeing but always managed to end up at some point in the bars in Yongsan and Iteawon, which are suburbs of Seoul. Seoul was the 8th Army's territory, and they were not under the same curfew and control restriction that we were up at Camp Pelham. Blowing off steam and staying up until the wee hours seemed like such an indulgence.

My last duty at Camp Pelham was guard duty when President Ronald Reagan visited the DMZ on the 13th of November. He would be the first American President to visit the DMZ. That was a tremendously bold statement given the current environment.

Within the past few months the Soviets had shot down a Korean jetliner killing 269 passengers, 241 service members were killed when the Marine barracks in Beirut, Lebanon was bombed, Marines had just invaded the Caribbean Island of Grenada held by Cuban forces, and there was the assassination attempt on the South Korean President.

I was so short, only a few days from heading to the Turtle Farm. I was so short my fingers were touching! It didn't matter. The President of the United States was at the DMZ and needless to say the whole area was extremely tense. For weeks the area had been abuzz with military vehicles massing in case of a problem.

To make matters worse, two South Korean soldiers had deserted their guard post outside of Seoul on the 7th of November and had shot and killed another guard. They were at large with M16 rifles, four grenades, and 600 rounds of ammunition. After killing two more people they finally committed suicide the next day when they were surrounded while holed up in a building.

The normal guards that patrolled around Camp Pelham were civilian contractors. They patrolled the perimeter and manned the guard towers strategically placed around the camp. They were not a slack bunch, and you didn't want to try and sneak into the camp after curfew because they would shoot you. This was definitely a "shoot first and ask questions later" type of place. But on this occasion, they wanted Americans standing guard until the president had cleared the area.

Camp Pelham Civilian Guard

I was assigned guard duty at night and stationed in one of the guard towers. It was the middle of November and freezing cold. I was issued live ammunition and told to climb up the tower and radio if I saw anything unusual. It was cold as hell, and I was totally bundled up in all my cold weather gear. The tower was 30-40 feet high, open to the elements, and was made for one person. The ladder up to it was narrow and straight up. It was a tough climb up with my bulky clothing, radio, and weapon. Once up there, for the first time since I had been at Camp Pelham, I had a bird's eye view of the sleepy little village of Sunyu Ri. It was late at night, after curfew, and the village was at peace. It was a nice final impression of my neighbors for the past year.

President Reagan on the DMZ

President Chun of South Korea said he had ordered forces to be ready to fire an artillery barrage in between President Reagan and the North Koreans in case he was attacked and said President Reagan was the only leader that had the courage, fortitude, and leadership to make such a visit. President Reagan came to the most hostile place in the communist world and from guard post Collier stared over the 1,100 yards of barbed wire and mine fields that separated him from North Korea.

President Reagan later gave a speech in the mortar bunker at Camp Liberty Bell, which is located near guard post Collier. In his speech he said, "Somebody asked me if I'd be safe up here so close to North Korean troops, and I said, I'll be with the 2nd

Infantry Division!" Such a statement from such a great man made us all very proud.

I received my orders to Fort Polk, Louisiana and was to leave Korea on the 18th of November, exactly one year from my arrival. I didn't ask to go to Fort Polk, it was chosen for me by the Army. I didn't mind, John Fisher, Lewis Blankenship, Terry Bear, Steve Wiliamson, and several of my other friends had received orders there also. So, it was no big deal. Tim Bailey and a few others had received orders to Fort Riley, Kansas. And Mike Guertin had asked to stay another year in Korea.

I had been waiting for my orders like a kid waits for Santa on Christmas morning. I was out in the village and it was getting late and my buddies and I decided to head back to camp before it got close to curfew. We stopped at one of the street venders selling fried potatoes. The way they served them was by taking a piece of paper, making a cone out of it, and serving your fries in the cone. The paper was always recycled, and I don't mean recycled like now-a-days. It was always trash paper from the camp. As I was eating my fries, I noticed the paper had typewritten English on it so I gave it a look. It was a copy of my orders! Now what were the odds of that happening? Not my actual correct orders but obviously a draft copy with errors that had been thrown in the trash. I couldn't believe it. The next day I went to the company clerk and asked him if he had my orders, and he did. I told him that I found a copy in town wrapped around my fries and he calmly said, "It happens."

Before leaving Camp Pelham I was awarded the Army Achievement medal for "exceptionally commendable service."

It was a total surprise. I totally credit our young 2nd Lieutenant, Paul W. Kelly, who saw enough in me to put me in for the award. And my rack mate SGT Joo Myung Hoo presented me with a plaque, which really meant a lot to me. It said, "Two friends who lived and worked together, may we never forget each other and the good times we had."

After checking out of Camp Pelham I headed back to the Turtle Farm to out-process.

My Grandfather, Ben Churilla died of cancer, and Cousin, Tim Semler, was killed in a house fire while I was in Korea. I didn't find out until weeks later in both cases. Phone calls were very hard to arrange at that time and the only means of communicating back home was via letter. My folks were still living in Australia and that made it even harder. I never made a phone call while in Korea and my only communication with my family had been via letter.

Once finished with the Turtle Farm I boarded a civilian plane that would take me home on leave and then to Fort Polk, Louisiana. As I made myself comfortable for the long flight home, I never thought of KAL 007 or anything else that happened over the past year. My only thoughts were for the moment and leaving Korea. I was sitting back in the smoking section and after taking off I lit a cigarette and took a nice long drag on it. As the stewardess walked past me, she asked if I would like something to drink and I said "Yes, a beer please." The stewardess asked me if I was 21 and I replied, "No I'm 19." She flatly told me I wasn't old enough to drink!

FORT POLK LOUISIANA
1st PLATOON, E COMPANY, 7th ENGINEER BATTALION,
5th INFANTRY DIVISION (MECHANIZED)
18 December 1983 – 16 June 1985

I spent most of my 30 days of leave between the Semler farm in Pennsylvania and my parents' house in Manassas, Virginia where they had recently relocated after leaving Australia. When it was time to leave, I caught a flight from Virginia to New Orleans, Louisiana. After spending a few days visiting with Steve Williamson, a buddy from Korea who lived in New Orleans, I caught a Greyhound bus up to Fort Polk. It was my first time in the deep South and the five-hour bus trip gave me a chance to see the southern countryside while stopping in numerous small towns to transfer passengers and mail.

There was a marine on the bus who was also headed to Fort Polk, or home on leave. He looked so sharp in his dress uniform compared to mine. It made me think back to when I was at the Marine Corps Recruiters office and could only knock out one pull-up. I'm sure I could do a little better now, but probably still couldn't do 5.

Fort Polk is in the center of the state toward the western border with Texas. The fort was named in honor of Leonidas Polk, the first Episcopal Bishop of the Diocese of Louisiana, and a general for the Confederacy during the Civil War. He was killed in action on June 14th, 1864, while scouting enemy positions near Marietta, Georgia.

Now this is the second Confederate named installation in a row I had been assigned to. I had nothing against the Confederacy, except for that fact that my great-great grandfather, John Gunsallus, had fought for the Union during the entire civil war with the 51st Pennsylvania Infantry. He entered as a private in 1861 and was mustered out as a first lieutenant in 1865. He saw action at Roanoke Island, Newbern, Camden, 2nd Bull Run, Chantilly, South Mountain, Antietam, Fredericksburg, Campaign of the Mississippi, and Spotsylvania Court House just to name a few. Two of his brothers, Zachery and Samuel, also fought for the Union. Zachery was with the 13th Pennsylvania Calvary and Samuel was with the 148th Pennsylvania Infantry. Ironically all three brothers were fighting around the same location in Spotsylvania, Virginia when Samuel, who had survived the battle of the Wheat Field at Gettysburg, was killed at the battle of the Spotsylvania Court House on May 16th, 1864.

Now at Leonidas Polk's namesake I was assigned to another in-processing center. Fort Polk was an old fort, and I was once again assigned in a WWII era barrack. These buildings were typical two-story buildings made of wood, with no apparent insulation, and a white paint exterior which seemed to be peeling away from exposure to the hot southern summers. I was assigned to the second floor and in a room with one other guy.

While here, I had $20 stolen from my locker. I suspected my roommate. It was a little unsettling. In Korea, slicky boy was always a threat, but I never worried about my Army buddies stealing anything from me.

Even though my roommate was in the Army, he had no special connection to me, he didn't care about me. He had no reason to. I was just another person to him. And that is the huge difference between someone you have gone through hardships with, someone you have entrusted your life with, and someone you haven't. That's the big difference with being in a squad, section or platoon with someone verse being in the same division as someone. The trust and respect dissipate the further you get from day to day routines together.

My stay at the in-processing center was only a few days and then I was transferred to my new unit Echo Company, who were also known as the "River Rats." Echo Company belonged to the 7th Engineering Brigade of the "Red Diamond" 5th Infantry Division consisting of five officers and 146 enlisted men. It was almost a mirror image of personnel and equipment that I had left in Korea.

The Commanding Officer of the company, Lieutenant Greg Kniesler even had some nifty business cards made up for our bridging service. He also had a very informative booklet entitled "River Crossing Made Easy" published that explained in detail the organizational make up and function of a Ribbon Bridge Company.

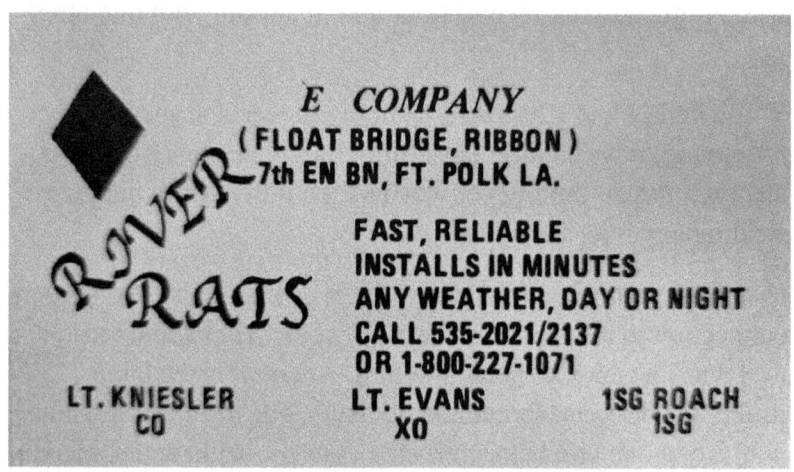

Echo Company Business Card

My new accommodations were a night and day difference from what I had experienced in the Army to this point. I arrived to new state of the art digs. The barracks and surrounding buildings such as the mess hall, barbershop, and orderly room were all new, modern, and made of solid brick. The barracks were three stories high and had four rooms on each floor with a recreational area in the middle which had a sofa, table, and several lounge chairs. Each room accommodated three people. Each person had their individual wooden bed with a nice large wooden wall locker and desk. There was also a shower, sink, and toilet in each room. This was like five star living compared to Camp Pelham!

I should probably only have given it four stars just because of the huge cockroaches that were all over the place. It was typical for Louisiana, but not for this Northern boy. These things were everywhere and there was no use killing one because five more would take its place. It was a shame to be lying in such a nice bed in such a nice room and have these two-inch-long things crawling over your blanket and up the wall.

Fort Polk Barracks

I moved around several times in the barracks but spent most of my time on the second floor of the building that housed our platoon. My roommates were McDonald from Philadelphia and Tate from Chicago. That's the way it was in the service, you always addressed a guy by his last name, and you always knew where he came from because that's all they ever talked about. Tate and I had actually been in Korea together but in different platoons, so we didn't run in the same circle of friends over there.

I was joined by other guys I had served with in Korea like Lewis Blankenship, John Fisher, Terry Bear, Avery McGee, James Tate, Steve Williamson, and Milo Van Auken. It was good to be around friends again. I arrived just before Christmas like I did in Korea and would spend my second year in a row away from home for the holidays. It wasn't too bad because the fort was pretty much shut down for the holidays. Most of the company was on leave and it gave me time to get acclimated to my surroundings. The Army always took care of you, and I got a knock on the door on Christmas day with people bringing us care packages and making sure we were doing okay over the holidays.

Our daily routine at the fort was similar to that of Camp Pelham. Up at 0500 for PT and a several mile run, followed by chow and muster down at the motor pool. We ran in government issued banana yellow sweat suits here and the fort was so big we could run for miles and miles and never even get close to running out of room. The day at the motor pool consisted of maintenance on truck, boat, and bridge equipment. This was another ribbon bridge company like in Korea, so it only took a few weeks for me to come up to speed with their routine.

But unlike Camp Pelham, things at Fort Polk were spread out. The barracks, barbershop, mini PX, Dining Hall and company building were all relatively close to each other. But the motor pool was over a mile away. And if you wanted to go into the nearby town, you needed a vehicle. It was too far to walk.

It seemed like we were always going out to the field at Fort Polk compared to Korea. Going "out to the field" was our term for what the Army called field training exercises (FTX). These usually consisted of a week of training and roughing it out in the Louisiana scrub and swamps. Fort Polk covered over 100,000 acres so there was plenty of room to roam.

The climate and terrain sort of resembled Vietnam and it was used to train soldiers heading to fight there during the war. It even had a mockup Vietnamese village, which was known as "Tiger Land."

Being out in the field was nasty. You never changed your clothes, didn't wash, did your business in holes you dug, and stunk. We were so nasty after an FTX that we had a rule that when we came back to the barracks no one was allowed in the room with their clothes or gear until they had been washed.

The back woods and swamps of Louisiana are a nasty place to be roughing it. And it was easy to see why they used it to resemble Vietnam. The state bird was jokingly called the mosquito, and there were all sorts of spiders and snakes that could really put you in a bad way. When we got to our field location, we camouflaged our trucks with netting and then dug our foxholes. The foxholes were always filling with water because the water table was so high there in Louisiana. Once you got one dug you seemed to spend the rest of the day bailing it out.

After we got all that done it was time to pitch our little two-man tent for shelter. Now this is when you had to really trust who was assigned to be your foxhole buddy because each of us had half the tent or was supposed to. When you are issued your field gear you get one half of the tent canvas, four tent stakes, one piece of rope, and one tent post which comes in three pieces. Now to put the tent together your foxhole buddy is supposed to have the same items you do so that together you have the makings for one tent.

Going to the bathroom was always an adventure. There were no portable toilets and you had to go dig a hole if you needed to go. The deal was you dug it, used it, and most importantly fill it. Going really wasn't an issue because the "C" ration meals we would eat would bind you up for a week, just about how long we were in the field. The "C" rations were canned food meals nicely packed into a small brown cardboard box.

Getting your meal was the luck of the draw. Breakfast was usually trucked out to us in these huge stay warm containers which didn't keep things warm. You would go through the chow line, get your breakfast, fill your helmet with hot water for shaving and washing up from a huge drum of hot water, and pick out your two "C" ration meals for your lunch and dinner. There were only about six varieties of "C" rations that I ever saw, and the best ones went fast. Once back at the foxhole you ate your cold breakfast and washed your face and shaved in your helmet of water. After that you commenced to trade for better "C" rations.

The worst one was ham and eggs, followed closely by chicken loaf. It really didn't get too much better, but I was fond of the classic beans and franks. These "C" rations had a shelf-life of at least 20 years so you can imagine what ham and eggs looked and smelled like when you opened the can. Every meal had the entree in a can, a dessert such as canned fruit cocktail, a can of a bread item such as crackers or cookies with a cocoa powder packet, salt, pepper, matches, toilet paper, a spoon, sugar, instant coffee and gum. Some of the older ones even had cigarettes. These last items came in paper packets and were tough to keep dry in the ever damp bayous of Louisiana. Socks, underwear, and toilet paper were the items you always tried to keep dry no matter what.

Everyone carried a small can opener known as a P-38 to gain entry into their canned "C" ration cuisine. Most of the time these were eaten cold, but every now and then I would build a little fire at the bottom of the foxhole, so the flames could not be seen, and heat them up. It didn't make them taste any better, but eating them warm made them go down a little easier. I also really enjoyed the cocoa powder and always tried to trade for it. By just adding a little water to it you could make a thick chocolate paste just like pudding.

On one FTX in June of 1984 we left Fort Polk via convoy to an area in Louisiana on the Red River. It would be about a 60-mile trip from Fort Polk. This was unusual and one of the only times we left Fort Polk on maneuvers. Our FTXs were usually in the vicinity of this little lake, appropriately named Engineer Lake, which we used for bridge ferry training. Driving out on the highways and byways of Louisiana was exciting and I felt liberated from the fort. But there are always risks when you get military vehicles out on open highways and 2 civilians were critically injured when they collided with M60 tanks on two separate occasions.

This FTX turned out to be a major evolution dubbed operation "Devils Fury Four" consisting of over 7,000 soldiers and 600 armored vehicles. The objective was to try and cross the entire 5th Infantry Division in one day. A huge undertaking. It was one of the largest field maneuvers conducted at Fort Polk and the first this large since World War II.

After a good day of driving, we wheeled into a huge cow pasture that bordered a part of the Red River. It was a typical pasture with knee high grass and cows. Like always when out on a FTX, we conducted ourselves as if we were in the middle of an active war zone. That meant full field dress, 24-hour watches, foxholes, "C" rations, and everything was camouflaged. So, as we came into the cow pasture we set up toward the river and commenced to camouflage everything. We pitched our two-man tents and would have dug foxholes if it were not for the fact the cow pasture owner was not keen on the idea. That night when the sun set, the cows were grazing way over on the other side of the pasture, about 400 yards away. When the sun rose, they were grazing all mingled in with our equipment and tents. It was actually pretty cool to herd them along their way and back out into open pasture.

That first night also revealed our first casualty. One of the guys was bitten by a spider, or something, and had broken out with a rash from head to toe. I mean he looked like hell. This was no small skin irritation, it was full blown head to toe red bumps. No one wanted anything to do with what had gotten hold of him! He was transported back to the hospital on the fort. It wasn't the first time that had happened. On another FTX a guy was bitten on the face and half his head swelled up. Southern Louisiana was known for having lots of chiggers, lice, Brown Recluse & Black Widow spiders.

AVLB Getting ready to Launch a Section of Bridge

I don't really remember conducting any bridging operations on the Red River, because it was unusually low. I think most of the crossings were conducted by our sister unit of Armored Vehicle Launched Bridge AVLB, which deploys a small bridge from an M60 tank.

But I do remember crossing the river. I was assigned with a couple other guys to take one of the inflatable boats and paddle over to the other side and scout for the enemy. We were not actively in a war game scenario with anyone else so there wasn't any real enemy, just us doing our normal and annoying pretending. So, the four of us climb into this black reconnaissance raft with our M16s slung over our bodies, loaded down with field gear, and paddled about 100 yards over to the other side of the river.

The river was moving pretty fast, and it took everything we had to paddle over to the other side. Once there we scouted around and as expected, found no enemy. We had gathered at a point just on the edge of the river bank behind a huge log. While lying there in the weeds, we discussed if we had spent enough time messing around to satisfy the command. One of the guys pulled out a joint and suggested it was a good time to get high, so we did. Mistake!

On the way back we were about midway across the river when we somehow flipped over. I think we started to get turned around, the current was moving too fast, and the raft just flipped on us. We all went into the muddy water along with our gear. It sure as heck scared any buzz I had out of me, and I was relieved to reach the shore. Luckily, we all made it to shore, which was amazing since we were all stoned and the water was moving so fast. Once there it was discovered that one of the guys had lost his M16 in the river.

Now this was a major problem because the only thing worse than getting bitten by something out in the field and breaking out into some disfiguring rash, was losing your weapon. You never wanted to lose your weapon and you never wanted your command to have to write a report on it. When a weapon is lost in the field everyone goes into lock down and your every action is all about finding the missing weapon. A mound of paperwork has to be filled out and you're the center of attention, exactly what you don't want.

So, the whole company spent the next day or so trying to retrieve the M16 via grappling hooks and anything else we could think of to no avail. At some point it was given up as lost and, with no M16 found, we finished our FTX and headed back to the fort. Although I'm sure the guy that was missing the M-16 still had plenty of long days dealing with the consequences of losing his weapon.

We usually smoked pot while out in the field. At that time in the Army there really wasn't a big penalty for smoking pot, usually extra duty. Harder drugs would get you discharged, and I never encountered them. We were all young, single, didn't have anything to lose, and the Army seldom conducted drug testing, so we never worried about it.

On another notable FTX we headed out into the scrub of Fort Polk. On this particular occasion we would be playing war games with other units in the area. We set up our perimeter in our normal location near Engineer Lake and dug in. On this FTX we were using the same MILES laser gun technology I had used during Team Spirit in Korea. My foxhole buddy and I had dug a nice sized hole and started to go out into the firing zone, out in front of our foxhole, to set up some trip wires. These were pretty cool. It was a wire that you strung across a path, or clearing, and the wire was attached to something similar to a bottle rocket. If someone tried to sneak up on us they would hopefully set one off and alert us.

It was cool and rainy out in the scrub, and it didn't take long for all of our stuff to get damp and wet. As usual our foxhole kept filling with water and needed constant bailing out. I remember one night sitting at the edge of my foxhole, against a tree, trying to stay dry with my rain gear on and poncho covering me. I was tired, cold, hungry, and I stunk. The only joy sitting out there was getting a good drag on a dry cigarette. The next good feeling was getting my turn to sleep and crawling into my nasty sleeping bag with all my clothes on for 55 minutes of sleep before I had to watch the foxhole for another hour.

On the second or third night we had heard that we might be attacked so we were all up and in our foxholes. Every now and again our squad leader would come crawling by to check on us and pass any information. At some point in the night a trip flare went off and it was on! The pitch black night was broken by the flashes of bottle rockets and there was the sound of MI6's firing and laser gear buzzing. Your adrenalin just about makes your head pop off at times like that. I can't imagine how it would be if it was actually life or death combat. As the battle heated up you could hear the platoon's M60 machine gun buzzing away and hear breaches in our perimeter. As quick as it had started, it ended with the assaulting force drifting back into the pitch black woods. After that we were all on pins and needles for the rest of the night.

The next night would be our turn. We started out in the afternoon and headed through the woods in a long sweeping motion toward the enemy. We wanted to take it very slow and get into position before dark. That way we wouldn't have to make a lot of noise stumbling through the woods at night. As usual, it was raining, and the dampness seemed to penetrate every layer I had on. We walked very slowly for several hours without talking. Every now and then we would get the motion to stop, and we would crouch down by a tree and cover ourselves with our poncho and have a smoke. That smoke always seemed to give me such a moment of relaxation and enjoyment in what seemed to be a miserable environment.

Eventually we made it to the location where we would hold up until night. As night fell, we waited a few hours hoping the enemy was getting tired and thinking we were not coming. Eventually we started to move forward, and someone sent up a trip flare and it was on again! Just like the night before the adrenaline started pumping and there was an escalation of gun fire, flashes of light, and the sound of laser gear beeping. I started running up this grade and toward a foxhole in front of me. I could see pretty clearly because the trip flares had really lit the place up. When I reached the foxhole no one was there and I hopped in and kept firing. I soon thought it best to get back to our regrouping location, so I headed back. On the way I ran into my fellow squad mates and we started howling and laughing about the whole event. I wondered, what happened to the guys in the foxhole I had hopped into? Had they been that scared that they took off, or was there never anyone there? Anyway, that was a memorable FTX and we swapped war stories for days after that.

In early summer of 1984 I was wondering why I had not been promoted with my peers to specialist four (SP4). When I asked my chain of command, they gave me the usual spiel that it should be coming down soon. After several weeks of no promotion, I began pestering my command again and they eventually sent me up to headquarters to see what was going on. Headquarters was a huge building in the center of Fort Polk that housed the clerks I needed to see along with the fort commander, a general. I had never seen anyone of a higher rank than a captain before, so I was a little more than hesitant to go up to headquarters where the general hung out.

There was just so much protocol with saluting, and what if the general's car drove by? I knew you were supposed to stop and salute the car, I think? I was having flash backs of getting my ass chewed out in front of the PX back in basic training. All of this was racing through my mind as I walked to the center of Fort Polk.

As I approached the building, low and behold there was the general's car parked out front with his red flags with white stars sticking up from the front bumper. That meant, in my mind, I could pass him at any moment! I treaded softly into the building and found my way to the administration section housing the clerks. Once there my clerk sat me down and opened up my very thin personnel record. He calmly said that the reason I had not been promoted was because the only entry in my record was that I had arrived at Fort Leonard Wood for boot camp. Now this was in the days before computers, and everything was hand written into your record. The clerk said he would update it and I should be promoted soon. True to his word I was promoted on the 1st of August to SP4. Oh, and I never did run into the general!

One of the duties I had to stand was charge of quarters (CQ) watch. This was a 24-hour watch stood at the company's orderly room. The responsibilities included answering phone calls and checking guys in and out of leave. One of the unusual responsibilities was administering medication to guys not responsible enough to be in possession of their own medications. One of those medications was Antabuse, which was a treatment for alcoholism.

SFC Manis – standing on the right

While in Korea I had served with a legend of a sergeant first class by the name of Manis. He was a John Wayne type of guy and a legend for being a great platoon leader and a heavy drinker. He was my platoon leader before SFC Garrett in Korea, but here at Fort Polk he was in charge of another platoon. Over in Korea and here at Fort Polk he was highly admired. He was so admired that he was simply known as "Warrior."

He was already old and beat up when I first got to work for him in Korea. Army life had taken its toll on Warrior. He wasn't alone, very few of the older sergeants were in good shape.

When we would form up for PT in the mornings they would be there to check in when their name was called for muster. After muster all those with medical profiles would be released and they would head back to the barracks before we would start to exercise. The rest of us would assume the front-leaning rest position and start doing push-ups! No one held it against them. Most of them were Vietnam veterans and had our respect.

So, on one of my CQ duty days I was told that SFC Manis would be coming in to take his Antabuse. I was to watch him take his pill and log in the logbook that he indeed took his Antabuse. The Warrior came in as expected and I handed him his pill, which he balled up in his weathered bear paw of a hand. He turned away and started to walk out of the orderly room. I called to him, "SFC Manis, I was told that I needed to see you take your pill." His reply was, "You did." And he walked on out the door. I should have never questioned him. I simply wrote in the log that SFC Manis reported as directed and had taken his Antabuse pill.

In the late summer of 1984, we were told that we would be heading to Germany for almost two months to participate in the European war games known as Return of Forces to Germany (REFORGER). We would be taking part in Operation Certain Fury while elements of the 2nd Armored Division - with elements coming from Fort Hood, Texas - and British troops would be participating in Operation Spearpoint.

The 5th Infantry Division would be sending two brigades, consisting of about 9,000 soldiers. It would be the first time they had participated in the war games since 1978.

We were all excited about getting out of the backwoods of Louisiana and going to Germany! A lot of the guys in the company had done tours in Germany, so the stories of beer houses and Oktoberfest started to season the excitement of deploying. In preparation we took classes on the German culture, had to pass European driving tests, and were briefed on the possibility of espionage that came along with the REFORGER event.

The Russians were highly interested in what was going during REFORGER and had intelligence gatherers out trying to gather information on NATO equipment and movements. This was during the Cold War when Germany was still divided between a communist East Germany and democratic West Germany. The Berlin Wall which separated East and West Berlin would not come down for another five years, in November of 1989, and Germany would not reunite until 1990.

We were told to report any suspicious activities we saw and disrupt the efforts of suspected intelligence gatherers if we could. The theory of disruption would go down something like this; we see a car full of seedy looking guys pull up next to the convoy as we are driving down the road and they start taking a lot of pictures. We were supposed to maneuver our vehicles to box in the suspect vehicle and trap them until authorities could get there. Sounded like a James Bond movie to me!

The plan for going to REFORGER was to leave all of our bridge equipment in Louisiana and pick up bridge equipment when we arrived in Germany. Other units of the 5th Infantry division shipped about 800 vehicles from Fort Polk. They would begin rolling off of ships in Antwerp, Belgium around the 11th of September. On the 8th of September we loaded up in Air Force C-141 troop carrying aircraft and flew out of Barksdale AFB, located in northern Louisiana, to Rhine-Main airport in Frankfurt, Germany. We boarded the C-141 aircraft wearing our military BDUs and carrying our weapons. Our duffle bag with all of our other gear was loaded as checked baggage.

We had a packing list of items that we had to bring in our duffle bag and our squad leader inspected us to ensure we had them. The last thing they wanted to have to deal with was someone leaving their tent half our cold weather gear behind. And since we all had the exact same duffle bag, we painted our unit and the last four numbers of our social security number on the bottom of our bag. That way when they were all stacked up like cord-wood we could easily identify our individual bag.

When we landed in Frankfurt it was cool & raining. I mention this because it seemed like it rained or was overcast during our entire deployment. From Frankfurt we boarded buses and drove about 100 miles southwest to the town of Pirmasens, which is about 5 miles from the French border. The drive was beautiful and so pristine. The windows were a bit foggy, and the rain falling and being whipped up from the bus seemed to me what Europe should feel like for some reason. We drove on the autobahn and through little villages that had been there for thousands of years. We arrived at the U.S. military instillation named Husterhoeh Kaserne on the northern edge of the city of Pirmasens and would spend the next week there drawing our bridge equipment out of storage.

Husterhoeh Kaserne was part of a huge network of storage facilities in Germany known as Prepositioning of Material Configured in Unit Sets (POMCUS). The facilities housed equipment staged in the event of a war with the Soviet Union. They had been positioning equipment here since the Berlin Crisis in 1961. Everything was there except the soldiers who would be flown over to Germany to man it.

The U.S. storage facility at Husterhoeh Kaserne was unbelievable. There were these huge aircraft hangar sized buildings that stored all the bridge trucks, boats, and equipment we needed. They also housed tanks and APCs being drawn out by other units. Some of these facilities were even humidity controlled to prevent the tires on vehicles from dry rotting. All this equipment is stored & maintained by the U.S. Army Combat Equipment Group Europe (CEGE). It's estimated that they had over 20,000 wheeled & tracked vehicles in storage throughout Germany & Belgium.

The process took about a week because this equipment was in what they called long term storage. Everything needed to be mechanically readied, such as having oil and fluids added, tires checked, engines started, and vehicles inspected. The military facility there was enormous and had all the amenities we needed, along with German merchants selling food. I was astonished at how many different varieties of bratwurst you could get from the guy at the bratwurst cart!

Once we had drawn all of our equipment and had it inspected for release, we were ready to roll. The plan was to convoy further south east about 150 miles to a place 50 miles east of Stuttgart. From that location we would wait for the

REFORGER war games to start. Traveling 150 miles in the states doesn't seem like a big deal, but in Germany it would take us about a week. We had to convoy our five-ton bridge trucks, which are huge, through country roads and towns only meant for small car traffic. The height of our equipment would also make it very hard to go through the small villages and towns, so we would need to by-pass them.

It was a stop, wait, and go process of driving a few miles then stopping to wait for the scout jeep to see if we could make it down a road or through a village. Instead of trying to drive a whole convoy through the countryside at once we were broken up into smaller groups of about five vehicles and would stagger our groups by several miles. This would ease the disruption to the locals.

Our biggest obstacle turned out to be the weather and not the roads. This would turn out to be the wettest September in southern Germany in 40 years. Not only was it damp but this was the start of winter in Germany, and it was cold. I put on my long underwear, field uniform, and rain gear complete with rain boots in early September and did not take it off until we finished in late October.

The rainy weather made driving more difficult. The roads were so narrow that if you got off them you had a good chance of getting stuck in the mud of the fields that came right to the road's edge. We would transit during the day and then regroup at night at a predetermined location and make camp, usually in some field or in the woods. This was like being back on an FTX in Louisiana, and for the most part there were no showers, or hot meals, just "C" rations.

It actually turned out to be a great way to really see the countryside and interact with the German citizens. As we would be stopped along some road or in a village we would get out and stretch our legs and mingle with the locals. Like I said some of the guys had been stationed here before and knew some basic German which always broke the ice.

On one occasion we had stopped just inside a small village and were told to wait. We knew there was a bigger town ahead and it would be the usual hour or so before the scout jeep had figured out a way through or around it, so we got out stretching our legs. Apparently, we had stopped right next to a restaurant and hotel known as a guest house. At the urging of those guys who had been to Germany before, we decided to go into the guest house and get something to eat. The Germans were very friendly and sat us in a nice big corner table. We had all of our field gear, along with our weapons, and must have been quite a sight. The locals made us feel right at home and we shed our winter jackets and piled our M16s in the corner. After eating a great meal, we thanked our wonderful hosts and went back to our trucks to await the order to move ahead. From that point on we always took advantage of any opportunity to eat and mingle with our very hospitable Germans hosts.

It wasn't unusual to get lost and detached from the main company while transiting through these cities and towns. It seemed like we were always passing broken down vehicles from the guys that were pushing ahead of us. Since we were way back in the reserve, as a bridge company, we saw a lot of them and the destruction they caused on the roads and fields. I remember one of our small groups getting lost for several days. There were also the mishaps, such as hitting houses and tearing up pastures with our vehicles. The locals took it in stride and we were only held up long enough for the details to get worked out before moving on.

It was reported that there were almost 2,500 claims filed by German civilians for damage during the first half of REFORGER 84, totaling $46 million dollars. Some of the worst incidents were a fuel truck exploding and burning two houses and 2,500 gallons of fuel being spilt in Aidlingen.

Sometimes we would stop off at remote U.S. military bases, which were very small, and seemed to just blend in with the countryside. They were usually perched on some hill or other strategic location. These were always welcomed stops because we could shower and eat in their mess hall.

But for the most part we just pulled off the road and into the woods every night, using the tree cover as camouflage. Usually, the German forests were made up of really tall and established pine trees, that offered great cover with their huge canopies. They were very open for 20 or 30 feet before spreading their branches, making it easy to get our trucks in. Once parked, we slept in the cab of the truck or pitched our tents.

This lifestyle of eating and sleeping on the move caught up with me and I started having stomach problems along with diarrhea. I saw the medic about it, and he gave me some anti-diarrhea medicine which seemed to slow things down. But I was still uncomfortable most of the deployment and the 'C" ration toilet paper was having a tough time keeping up with me.

In the field during REFORGER. 1st platoon, 2nd section. Front left to right; PFC Melvin, PFC Gonzales, PFC Allen. Back left to right; Sgt Lee, SP4 Lagman, PFC McDowell, SP4 Semler, SGT Magee.

Once we did get down to our operating area for the war games we primarily stayed in a field near a small village. That's the field and my section in the picture above. There was a dirt road that led from the small town back into the field and like everything else in Germany at the time, it was water logged. About the only duty we had to perform was to stand guard up in the town overlooking the road entrance. When I had the duty, I would always take up a position crouched down under a store front doorway, safe from the rain, watching the traffic coming either up or down the road. The locals would walk right by me as if nothing was going on, even though I was dressed in field gear with my helmet and an M16.

Map of Germany

When the war games started, we were in the rear, north of the Danube River, which was the border between us, the Blue Forces, and the enemy orange forces.

The enemy, Orange Forces, consisted of the VII Corps, 3rd Infantry Division, 2nd Armored Cavalry Regiment, 6-52nd Air Defense Artillery, 7th Engineer Brigade, and the 385th Engineer Group.

The good guys, Blue Forces, consisted of the 1st Infantry Division FWD, 72nd Field Artillery Brigade, 5th Infantry Division, 2/75th Ranger Regiment, 372nd Engineer Group, and F Company 425th Infantry.

The Orange Forces were supposed to try and put in their own ribbon bridge on the 17th of September for an attack, but the wet weather made the river too dangerous for the heavy M1 and M60 tanks to cross. Instead, they used the existing bridges near the town of Dillingen to kick off their offensive. Eventually, on the 19th the river slowed up a bit and the 10th Engineers of the 3rd Infantry Division and the 385th Engineers of the Army Reserves put in their ribbon bridges at Dillingen and Petersworth.

As the Orange Forces poured over the Danube River the M-1 Abrams and older M60 tanks battled it out over the German terrain until around the 24th of September when Blue Forces started to counter attack.

We were called up from the rear to assist in crossing the Danube in the counterattack. Instead of putting in a full bridge, we put in a raft. As we were conducting raft crossings with Blue Forces, several Orange Force Cobra attack helicopters swooped down on us and attacked. After an initial pass, the war game umpires ruled that we had sustained damage to two of our bridge bay sections and they had to be removed. We changed them out and carried on with our mission.

Putting in a raft on the Danube River with a MKII Jet Boat

By the 26th, the 3rd Battalion, 77th Armor of the 5th Infantry Division was leading the charge and driving deep into Orange Forces territory. By the 29th, the Orange Forces had been driven back well across the Danube and Certain Fury was over.

The training we conducted in Germany was exciting because at Fort Polk all we had to work on was Engineer Lake, in the middle of the fort. It was still and there was never any current to deal with. Here we had actual rivers again like in Korea. The equipment was also newer. We had the older style 27' propeller boats back home and here in Germany we were issued the new MKII twin jet boats.

When the war games finished, we convoyed back up to Pirmasens to turn in our equipment at Husterhoeh Kaserne. This was another lengthy process which took about a week. All the equipment had to be power washed clean, parked, all the fluids drained, inspected, and stored for the next REFORGER or actual war. After passing our equipment turn-in inspections we were sent to a temporary holding camp to await transport back to Louisiana.

The holding camp was at Heidenheim, about 20 miles east of Stuttgart, and was a mass of these huge green circus sized tents, with row after row of green cots in them. No walls or lockers, just cots. The toilets were a bank of port-a-potties that were set up outside and adjacent to the rows of tents. At one end of the camp was the bath house, which was made of converted mobile homes, filled with showers and sinks and covered with a huge tent. So, you didn't lose your wallet or valuables you were handed a plastic bag to take your valuables in the shower with you.

The mess hall was under another huge tent by the bath house. There was everything you needed here at tent city, a post office, library, barbers, venders, phones, arcade, theatre, and finance center to cash checks. Probably the biggest entertainment was watching all the field mice running around. There must have been thousands of them, granted we were in their field.

Because of all the rain it was one huge mud puddle, and all the tents were joined by a maze of wooden pallets acting as sidewalks. You did not want to get off of those pallets as it was solid mud! I don't think I saw a sunny day in Germany; it was all rain and fog.

We would be here for about a week before we flew out. One day we were lying around on our cots, and I heard a gunshot outside. I ran out to see some guys huddled around the row of port-a-potties lined up in front of our tent. Apparently, a guy in another tent had blown his head off in one of the port-a-potties. His body was removed, and you could see the blood splattered around inside. Word was the guys in the company he was in were harassing him and it got to be too much for him. I was told he had been to the firing range recently and they suspected he had pocketed a few rounds waiting for just such an opportunity.

The suicide, surprisingly, had little effect on me. I know that sounds cold, but since the guy was not in my nucleus of friends, my platoon or company, I was able to move on from the incident pretty quickly.

There were also lighter moments at the holding camp. Our hosts wanted to keep us busy, since idle minds are the devil's workshop, and offered day and overnight tours which I took advantage of. Although rainy, the weather was actually mild during our stay at the holding camp, which made it easier to sight see.

We had been told to pack a set of civilian clothes and this would be the first chance I had to put them on since arriving. I took a tour bus to Nurnberg and did some sightseeing around the city, and on another trip, I took a Rhine River cruise which was very scenic. The boat had a nice eating area and the views of the castles perched up on the hills were breath taking.

Enjoying some sightseeing while in Germany for REFORGER 1984

In the last week of October, we were bused back to Frankfurt and boarded another plane for the ride back to Louisiana.

Once in Louisiana it was back to the humdrum life of the fort and guard duty. I was recognized as the outstanding soldier of the guard on the 30th of December 1984 and then the outstanding soldier of the month for the 7th Engineer Battalion for the month of January 1985. The outstanding guard award got me out of standing post duty, which was very nice! We had been detailed to pull guard duty over the Christmas holidays at various outposts on base. We would muster up for inspection and the soldier with the best looking uniform and gear would be relieved from duty for that day. The rest would be transported out to the middle of nowhere to stand guard duty.

And I mean the middle of nowhere! I was carted out to ranges all over the base and had to sit out there alone for four hours at a time in the freezing cold, until I was relieved. You wouldn't think Louisiana gets cold, but it does. And my feet were always the first thing to get cold. The thin army boots offered little protection from the cold.

It all seemed pointless to me. Once out at the guard post there was no way to contact anyone if something did happen and I didn't even have live ammunition to stop anyone. Every now and then a jeep with the guard duty supervisor would come out and check on me. And I don't mean to see if I was okay, I mean to check and make sure I was awake and offered the correct challenge.

The correct challenge was for me to yell, "Halt who goes there?" After the answer of let's say, "Lieutenant Smith," I was to say, "Advance to be recognized." At which point the lieutenant would approach and show me his identification card. He would then question me on my general orders. If I did it correctly, I didn't get an ass chewing.

After several days of freezing my butt off out at these deserted ranges I decided I was going to be selected as the "outstanding guard." So, I spit polished my boots, creased my uniform, and folded my poncho perfectly. The effort paid off and I was selected! My reward was that I didn't have to go out to the range that day.

The outstanding soldier for the battalion was a little tougher. For that I had to compete with other soldiers at formal interview boards, as well as present a good military appearance. I reported to a board of non-commissioned officers (NCOs) in my dress uniform. They asked me questions while I was at the position of attention, and they were seated at a table in front of me. The board asked me questions about general military knowledge such as what were my guard duty standing orders, who was the general of the fort, who was the commander and chief, and so on. They would also get into some pretty strange stuff, such as what was the ball on top of the flag pole called, and what was in it.

They asked other intellectually deep questions such as; if you walked halfway into the woods, how far would you have to go to get out? Oh, and the answer they wanted to hear about the flagpole was that the ball is called a "truck" and it contained a bullet. The bullet was to be used by the fort's commanding officer to shoot himself if the base was ever captured. How he was supposed to get up there and get it, I have no idea!

My appearance and answers were good enough for me to win and advance to the subsequent outstanding soldier of the 5th Infantry Division. The competition was tough, and I didn't make it. That was fine with me. It was fun up until that point but the higher I went in the process the more nervous I became, and that just put me out of my comfort zone.

I had an excellent squad leader by the name of SGT William E. Magee. He had a lot of faith in me and had recommended me to the E-5, sergeant, promotion board in November of 1984.

In his recommendation he said that "He (SP4 Semler) has performed excellent in every aspect of being a soldier. He has proven to be one of the best SP4's in the platoon. He has been acting Section Sergeant (for SGT Lee) and was able to keep the same high standards, which were set prior. He has also been left in charge of a critical bridge mission as a raft commander. SP4 Semler keeps a high standard with his uniform and his appearance."

I appeared before the board, which was similar to the outstanding soldier board, and passed. But the waiting list for advancement was long. I had 400 and some odd points and needed over 500 to guarantee me advancement.

Since I had passed the board, but was just waiting for more points, I was appointed to the rank of acting sergeant on the 25th of January 1985. This meant I could wear the E-5 rank of a sergeant but would not get paid for it.

The appointment was a huge jump in the pecking order of the platoon and in the Army in general. An E-5 was considered an NCO and as moving into the leadership ranks. You were looked upon differently, given special privileges, and a whole lot more responsibility. Of course, at my age, I was more interested in the privilege than the responsibility.

With this unofficial promotion I was moved into the barracks housing the NCOs. My roommate would be one of the sergeants in my platoon, my Section Sergeant, SGT Lee. He had a double room to himself, and I don't think he was too happy to give up his space to me, especially when I was just an acting sergeant. So, there was already tension there.

Then he noticed after I moved in that I had a bottle of lice shampoo in the shower. He was definitely not happy about that and let me know.

Somehow, probably from out in the field, I had picked up pubic lice, or what we called crabs. I had pulled night guard duty down in the motor pool and when I was finished, I started to notice them. As I mentioned before, nasty stuff and exotic bug bites were just a part of going out to the field for weeks at a time. Well, this was my first time getting them and they tore me up. I headed over to the aid station and they gave me some shampoo to get rid of them, which worked just fine. But like I said, SGT Lee was not impressed and thought I had just contaminated his room. Having no choice, he put up with me and we just gave each other space.

The critical bridge mission that SGT Magee was referring to in my recommendation letter to the E-5 board was at a recent FTX at Engineer Lake. We had been in the field a few days and then tasked with putting in a raft, at night. We left our camp site after dark and headed to Engineer Lake in the pitch black, with only our blackout lights to guide us. It was an unusually dark and cloudy night. Every now and then the cloud cover would clear for a few seconds and let a glow of moonlight peek through. But it was extremely difficult to see the vehicle in front of you as we made our way down dirt roads to the lake.

Once there, I was assigned as the raft commander and took charge of assembling the raft. It was an enormous amount of responsibility and I jumped at the task. I had previously done some empty raft commanding, but never responsible for crossing any equipment or people.

Our mission was too cross M60 tank and APC vehicles. These being heavy tracked vehicles made them more dangerous to cross via a raft. As I loaded them on the raft, I had to be extremely careful that they would not come on too fast and spit the raft right out from under themselves with their swirling tracks. They also had to come on perfectly straight because they could not turn once on the aluminum raft, or their steel tracks would cut into the decking. Weight was also critical. An M60 tank had to be placed perfectly in the middle. To far forward and the raft would nose dive as we crossed and to far back would cause the raft to flip over backwards. APC also had to be spaced correctly. The tanks and APC's had full crews, and their lives were in my hands.

Once on the raft and ready to go I had to signal, using a flashlight in each hand, to the boats secured to either side. They would speed up, slow down, or reverse depending on my commands. When we reached the other side of the lake it was critical that I had the raft in line with the beach ramp. This required a lot of small adjustments with the boats to get the raft lined up correctly. Once lined up it was once again critical to make sure the tracked vehicles didn't go too fast and spit the raft out from under them as they exited the raft, and to make sure the boats held the raft secure to the beach ramp. This scenario played out for hours as we crossed a battalion of tanks and APC's.

I'm not sure why I was given such a responsibility at night. But I did have over 2-1/2 years of Ribbon Bridge experience at this time. I know we were low on personnel and my squad and section sergeants were filling higher positions due to vacancies. And I had just escorted an individual to the fort hospital earlier in the day for some sort of field accident. Whatever the reason, I was very proud and honored that my platoon sergeant, SFC Dan Pearson had enough faith in me to entrust me with this huge responsibility.

He was a tough sergeant who had been to Vietnam and didn't put up with any shit. That made it even that much of an honor.

Even though I didn't own a vehicle the entire time I was in the Army I was stopped twice by the police, both times driving my buddy John Fisher's 1979 Ford Mustang.

The first time we were at a party at some off base housing complex. It was late afternoon and a buddy and I were hungry and wanted to get some food. So, we decided to go to this fast food restaurant which was a few miles away on the main drag leading into Fort Polk. Now we were pretty lit up at the time we decided to go, and I don't know why John even gave us the keys. But John and I had been friends a long time and I'm sure he didn't even think anything of it. In any case, neither of us could have been thinking too clearly at the time.

We got into the car, which was parked under an open sided carport supported by four posts. My buddy got into the driver's seat, but left his door open for some reason. As he started to back out, the open driver side door got caught on one of the car port posts and started to bend the door. I told my buddy that he was too drunk to drive and to let me drive, although I'm sure I was no better off than he was. We made it over to the restaurant without an incident, but I misjudged the entrance, and accidentally pulled into the bank parking lot right next to the restaurant. That didn't seem to be a big deal because they were only separated by a grass divider with a curb. I thought I could drive right over the divider, curb and all, and into the restaurant parking lot. Mistake!

The curb was a lot higher than I thought and the grass divider was on a downward grade. I subsequently got the Mustang stuck on the curb. My buddy and I got out of the Mustang and tried to get it off the curb by rocking it back and forth to no avail. By this time we had the attention of the restaurant's manager, who must have called the police because the sheriff showed up.

The sheriff reminded me of the character Jackie Gleason played in "Smokey and the Bandit," with the Smokey-the-Bear hat, tan uniform, southern drawl, and big attitude. My buddy had walked over to the restaurant and called back to John to let him know we had his car stuck on the curb and needed help. I was standing there next to the Mustang when the sheriff sauntered up with his mirrored sunglasses. In a somewhat sarcastic tone he drawled, "Now what do we have here?"

I explained that I had missed the turn into the restaurant and I was trying to cut across the divider. The sheriff said "I can see that, now give me your driver's license." When I provided him with my Pennsylvania driver's license, he commented more than asked, "You're not from around here now are ya?" I thought to myself, this is going to be a long night!

At about this time a vehicle full of guys from the party, including John, arrived on scene. I told the sheriff that my buddies and I could lift up the car and get it off the curb, and we would be on our way. The sheriff told me that I was going to have to pay $50.00 to have a tow truck lift it off. Now $50.00 was a huge amount of money to me back then, about a quarter of my take home pay, and I thought about arguing the point. But fortunately, I had sobered up a bit by now and realized that he was actually giving me a break by not hauling my drunken ass off to jail! So, $50.00 later, and a small amount of work by the tow truck driver, and the car was off the curb. John took possession of his Mustang and we were on our way back to Fort Polk and calling it a night.

The second time I was pulled over, I was leaving the town of Lake Charles, Louisiana. John, Cecil Cooper, and I had driven over to Lake Charles to hit the bars. This was after we had taken Cecil to get a tattoo in Leesville. We had done some heavy drinking and were heading back home. I was driving, John was in the passenger seat and Cecil was in the back seat. It was past midnight, the streets were empty, and I was on the outskirts of town when I saw the blue lights flashing in the rear view mirror.

As the policeman walked up to the Mustang, John and Cecil were sleeping. The policeman told me to get out of the vehicle. At about this time Cecil pushed open the back door on the passenger side and just started puking all over the place. The policeman, unfazed by Cecil puking, asked "Let me see your driver's license." I complied, also handing him my military identification, and he asked me "Do you know why I pulled you over?" I replied "No" and he said "I pulled you over for running a red light" I was thinking to myself; I didn't remember seeing a signal light! I know I had been drinking, but I didn't feel like I was intoxicated.

As the policeman was looking at my license, another police car came racing up and suddenly stopped right next us. It startled me, and I was thinking I was in big trouble. The newly arrived officer yelled from his car to his associate that there was a code something or other going on back in town. The officer that had pulled me over tossed my license and ID card back at me and ordered, "Wait here, I'll be back."

After sitting in the car for several long, tense, minutes I thought; really? Should I wait? John was still sleeping and Cecil had resumed his passed out position in the back seat. I was thinking; that policeman had to have gotten my name for sure. I saw him write something down. Oh, screw it! And I took off! I sweated it out for weeks afterward, but never heard a thing about it.

My last major evolution in the Army was a trip back to Fort Leonard Wood, Missouri for training in the spring of 1985. I was excited to be going back wearing sergeant stripes. Our company had been tasked with going to Fort Leonard Wood for two weeks of refresher Bailey Bridge training. Since I had really only built the pontoon type bridge my whole career, this would be fun and different.

As a matter of fact, I never built a Bailey Bridge during my entire three years in the Army, except at Fort Leonard Wood. I don't even think there was an active-duty 12C company that had them. We were all Ribbon Bridge companies. I did hear of guys going down to South America and installing them. But that was done in a humanitarian capacity, not in any type of conflict.

A Bailey Bridge is the steel beam bridge most recognizable in a lot of old World War II movies. It comes in sections and to build it you basically keep joining sections together and pushing the bridge out on rollers across whatever it is you're crossing. As long as you have more bridge weight on land, the bridge won't fall into the gap your crossing. It was originally made to be assembled, disassembled, and moved.

Assembling a Bailey Bridge

They were used by a lot of small town municipalities across America to cross creeks and small rivers and I have seen them in small towns like Urbana, Maryland and Renfrew, Pennsylvania.

The bus trip up to Fort Leonard Wood was really scenic and relaxing. We stopped at civilian restaurants, like the Ponderosa, for dinner and the tabs were covered by the Army. It was the first time I had eaten at a civilian restaurant on the military's dime. It was actually pretty interesting. We were issued government vouchers for a certain dollar amount and we could go in and order whatever we wanted, up to that dollar amount. It felt good to have some control and freedom of choice for a change.

Once at Fort Leonard Wood we were assigned to the good old fashioned WWII barracks again. These were the two-story wooden barracks with a pot belly stove in the middle of the room and open bathrooms on the end. Very rustic compared to our nice barracks back at Fort Polk.

The training was really fun as we built Bailey Bridges for two weeks. But unfortunate we had an accident. It wasn't major, but still a bit unsettling. One of the guys got his gloved hand pinched under a steel beam. We immediately lifted the beam, and he slid off his glove, dropping it to the ground, to reveal part of a finger missing. He picked up the glove and shook it, and out came the rest of his finger. We scooped it up and sent the two to the hospital.

At one point we had a slow period and I decided to walk over to the recruit training area and see my boot camp barracks, Bravo 1-2. I was a sergeant now and not afraid of the drill sergeant's wrath! As I neared the training barracks area, the memories of boot camp flooded back to me. As I passed recruits, they would stop suddenly and come to the position of parade rest as I passed, afraid I was a drill sergeant out on the prowl. I gave them a stern "Carry on," and kept on my way. The buildings and surrounding area were just as I had left them three years ago and it was like seeing an old friend. When I got close to my old barracks a rush of fear suddenly came over me. I was too afraid to actually go into the B-1-2 building! What if

I ran into one of my old drill sergeants? What if SSG Strickland was still there and recognized me!? I did an abrupt about face and scurried back to the safety of my WWII barracks!

When we returned to Fort Polk, my three-year enlistment was coming to an end, and I needed to decide if I was staying in the Army or getting out. The Army wanted me to stay in and had set up a meeting for me with an Army re-enlistment counselor, who obviously tried to get me to re-enlist. He even guaranteed me promotion to a full paid sergeant!

But I was ready to move on and wanted to give the civilian world a try. Most of my longtime friends like John Fisher & Lewis Blankenship had already been discharged. And everyone else in my circle of friends were planning on getting out. And all we could talk about was getting back home. Plus, I couldn't see myself going to the field for the next 17 years. I looked at my platoon sergeant and his peers, looking older than they really were, out in the field, roughing it month after month. And I just didn't want to do that.

All my Dad's stories had come to life for me and I was homesick. I declined re-enlistment and waited to be discharged.

In my last days at Fort Polk the 7th Engineer Battalion was gearing up to ship out to the National Training Center at Fort Irwin, California. This deployment was to provide realistic desert training. Unlike REFORGER, we would be taking a lot of our equipment with us. And to get it there it had to be loaded on flatbed rail cars to be hauled by train.

While the rest of the battalion was loading equipment on rail cars, I was assigned to supervise a trash detail that patrolled the fort. It was a week-long detail and was pretty easy. I rode around in this modified pickup truck that had steps on the back for two guys to stand on. I had a driver and two guys hanging on the back standing on those steps. As we slowly drove around the fort's main and back roads, the guys on the back would hop off and pick up trash.

Every now and then I would get a call on the radio that there was trash reported at some location and we would divert to that location and pick it up. These were usually high profile trash issues, like a soda can in front of the general's house. One such emergent call was to get to the PX because a skunk had been killed right in front of it and they needed the smelly thing removed ASAP! We diverted over there and scooped the smelly animal up, deposited it in a dumpster behind the PX, and continued on with our duty.

Before leaving the Army a friend of mine, Robert Beck, and I took a motorcycle trip over to Galveston, Texas. I must have been out of my mind riding on the back of that thing for over 500 miles there and back. It was probably the most dangerous thing I did in the Army! While in Galveston we got drunk and ended up sleeping on the beach. When we woke up in the morning, we dusted off the sand and started to look around for someplace to eat. We saw this GP-medium tent there on the beach and decided to see what was going on. GP-medium tent could mean Army guys, hey maybe they had food!

When we walked into the tent there were these guys in blue uniforms lounging around listening to short wave radios. I asked what unit they belonged to and they told me they were with the United States Coast Guard. I wondered; why didn't I join the Coast Guard? I could have been sitting here on the beach the past three years!

On my last day at Fort Polk, I waited at the bus stop for an evening Greyhound bus to New Orleans. Standing there by myself, with everything I owned stuffed into my green duffle bag. It seemed like such an anticlimactic end to the past three years. And I was basically right where I started off, unemployed and uncertain of the future.

The bus picked me up, took me to New Orleans, and I caught a plane home to Pittsburgh. I wasn't completely out of the Army yet. I still had 3 more years of inactive reserve time to complete. But this didn't require any type of mustering or reporting anywhere. It just meant that I could be called back up if needed.

AFTER THE ARMY

I arrived back home in Butler, Pennsylvania where my parents had settled down after my Dad retired from the CIA. My plan was to live with my folks for a few months until I could find a job, move out on my own, and resume civilian life. The town of Butler was reeling from the loss of one of its biggest employers, Pullman Standard. They made railroad cars and had recently shut down, flooding the area with the unemployed. The loss of Pullman also affected all the businesses that supported such a huge organization. Finding a full-time job on my own was proving to be nearly impossible.

I remembered when I was being discharged from the Army, I was told to register with my local unemployment office and that a veteran's representative would help me with finding a job. So, I headed down to the local office in Butler to see my representative. When I arrived, the place was packed with the unemployed. I waited my turn and filled out the required forms to enter me into the job pool. When it was my turn to see the veteran's representative, he didn't have any good news for me. As he took my paperwork and moved it to the bottom of this foot high stack on his desk he said, "My best advice to you is to go back into the service."

After three months of job hunting, working odd jobs, and not finding a full-time job, I was demoralizing. It was hard enough having to move back in with my parents who still had two of my school age siblings at home. I also found that in the civilian world I was on my own and it gave me an empty feeling. I missed the strong relationships I had with my Army buddies. That got me to thinking about going back into the military. I was still in the inactive reserve, but I didn't want to go back into the Army. And then I remembered those Coast Guard guys on Galveston beach.

So, while visiting my grandma and Uncle Mike on the farm I pulled out the phone book and looked up the Coast Guard. The nearest recruiter was in Monroeville, which was about 35 minutes away. I called them up and talked to a recruiter. After several questions about my discharge code from the Army it looked like I qualified and was asked to come down to the recruiting office. That would be the beginning of over 22 more years in military service.

My friendships in the Army have lasted through the years and I still get together every few years with some and stay in contact with others via social media, phone calls, and texts. It's funny how we all wanted to get as far away from the Army as possible when we were in, and now years later we all remember those days as some of the best in our lives.

I even made it back to Camp Pelham and Korea in 2016 and 2023. The camp is no longer there. It had undergone a name change in the early 1990's and was fully closed and handed over to the Koreans in 2004. As of my last visit in 2023, it was completely abandoned and awaiting redevelopment as a housing project.

The 2nd Infantry Division had vacated all of the small camps up near the DMZ, but still maintains a presence in Korea, just further south.

Fort Polk is still active as of 2023, but I have not made it over to see what it looks like these days. The 5th Infantry Division was unfortunately inactivated in 1992 and redesignated the 2nd Armored Division. You may remember we deployed to REFORGER 84 together. With this change came the move of personnel and equipment from Fort Polk to Fort Hood.

I hope you found this an interesting read and if you were in the service, it brought back some fond memories. If you're interested, I have multiple videos from my time in the Army and trips back to Korea on my YouTube channel. You can find the link in the next chapter.

ABOUT THE AUTHOR

Ed Semler retired from the United States Coast Guard in December of 2007 with over 25 years of military service in both the United States Army and United States Coast Guard. In the United States Army he was an enlisted man and was honorably discharged as a Specialist Four (E-4). While in the United States Coast Guard he was enlisted, obtaining the rank of Master Chief Petty Officer (E-9), was commissioned as an officer, and retired as a Lieutenant (O-3E).

Fully retired, he resides in Schulenburg, Texas with his wife Jana, a retired Air Force senior master sergeant. Please feel free to check out Ed's other books at www.edsemler.com email him at mkcm378@gmail.com and check out his YouTube channel www.youtube.com/@MKCMLT

His other publications are;

"Around The World," a memoir of his 25 years of service as an officer and enlisted man in the U.S. Army and U.S. Coast Guard

"U.S. Coast Guard Cutter Sherman (WHEC-720) Circumnavigation Deployment 2001" which details the

Sherman's historic circumnavigation of the globe and deployment to the Persian Gulf in 2001

"The Three Gunsallus Brothers" a story about fighting for Pennsylvania during the Civil War

"Sam Houston & Napoleon Bonaparte Meet On The Civil War Battlefield" a true story of the Walker brothers

"Thoughts On Being A Chief Petty Officer" a take on military leadership

"Fighting For Pennsylvania In The Early Years 1763 to 1783 – The Story of Captain Thomas Askey And Lieutenant Richard Gunsalus Of Cumberland County"

"Joe Semler Playing Baseball in the 1920's &30's"

"Alice Springs Australia Adventures In The 80's"

"Count On Us Coast Guard Cutter Dependable – Law Enforcement And Search & Rescue"

"United States Coast Guard Tragedies"

"In Their Own Words – Short Stories of Pennsylvanians in the Revolutionary War"

"American Sailors & Marines During The Revolutionary War"

www.ingramcontent.com/pod-product-compliance
Lightning Source LLC
Chambersburg PA
CBHW070549050426
42450CB00011B/2780